Ben Stacy Jerrik (Ed.)

NearMap

AF693900

Ben Stacy Jerrik (Ed.)

NearMap

Aerial Photography, OpenStreetMap, Digital elevation model, Ipswich, Queensland

Part Press

Imprint

Publisher:
Part Press is a trademark of
International Book Market Service Ltd., 17 Rue Meldrum, Beau Bassin, 1713-01 Mauritius
Email: info@bookmarketservice.com
Website: www.bookmarketservice.com

Published in 2012

Printed in: U.S.A., U.K., Germany. This book was not produced in Mauritius.

ISBN: 978-613-9-15237-7

Contents

Articles

References

NearMap

Type	Subsidiary, Proprietary company
Founder(s)	Stuart Nixon
Headquarters	Suite 8, 281 Hay Street, Subiaco, Western Australia, Australia
Area served	Worldwide (Aerial photography currently Australia only)
Key people	Simon Crowther, CEO
Owner	Ipernica Ltd
Website	[nearmap.com nearmap.com]
Alexa rank	▲ 129,553 (January 2012)[1]
Type of site	Web mapping
Advertising	None
Registration	Optional (Free users) Required (Paid users)
Available in	English
Launched	November 11, 2009
Current status	Active

NearMap Pty Ltd is an Australian provider of high resolution aerial imagery (and not satellite imaging).[2]

NearMap currently operates a web portal which serves up its imagery and terrain (DEM) tiles, along with OpenStreetMap map tiles, all three of which are free for the public and most commercial businesses to use. All government agencies and some commercial businesses need to apply for a licence to use the tiles.

The provider currently covers 2.26 per cent of Australia's landmass, with coverage primarily for capital cities and major regional towns. Special imagery may be flown for events such as natural disasters, for example, 2 cm resolution coverage of Brisbane and Ipswich during the 2010–2011 Queensland floods. Imagery is normally updated for 60% of Australia's population every month at a 7.5 cm (3.0 in) resolution and an additional 15% every three months at a 10 cm (3.9 in) resolution.

On the 21st of November 2008 Ipernica acquired NearMap for A$4 million plus shares and options along with A$1 million of investment.[3]

Free licencing

From their About page: *We specifically encourage creation of Creative Commons licensed information derived from our PhotoMap content, so that everyone can share and build a greater understanding of our planet.*[4]

The OpenStreetMap project previously used NearMap provided aerial imagery, with explicit support from the company, under the CC-BY-SA license, however it is no longer used due to OpenStreetMap licencing changes making NearMap imagery incompatible.[5]

References

[1] "Nearmap.com Site Info" (http://www.alexa.com/siteinfo/nearmap.com). Alexa Internet. . Retrieved 2012-01-02.

[2] (http://forum.nearmap.com/forum/viewforum.php?f=37)

[3] "Stuart Nixon's NearMap Acquired" (http://apb.directionsmag.com/archives/4545-Stuart-Nixons-NearMap-Acquired.html). . Retrieved 24 November 2009.

[4] "NearMap : About" (http://www.nearmap.com/nearmap/about.aspx). . Retrieved 24 November 2009.

[5] "NearMap Community" (http://www.nearmap.com/community/contribute.aspx). . Retrieved 24 November 2009.

External links

- Official Website (http://www.nearmap.com/)
- Official NearMap Twitter Account (http://twitter.com/nearmap)
- Official NearMap Facebook Page (http://facebook.com/nearmap)
- NearMap on the OpenStreetMap wiki

2010–2011_Queensland_floods

2010–2011 Queensland floods

A woman trapped on the roof of her car awaits rescue during the Toowoomba flash flood

Duration:	December 2010 – January 2011
Fatalities:	35 confirmed dead and 9 missing
Damages:	A$30 billion[1]
Areas affected:	Much of central and southern Queensland including Brisbane, Rockhampton, Emerald, Bundaberg, Dalby, Toowoomba, and Ipswich

A series of floods hit Australia, beginning in December 2010, primarily in the state of Queensland including its capital city, Brisbane. The floods forced the evacuation of thousands of people from towns and cities.[2] At least 70 towns and over 200,000 people were affected.[2] Damage initially was estimated at around A$1 billion.[3] The estimated reduction in Australia's GDP is about A$30 billion.[1]

Three-quarters of the state of Queensland was declared a disaster zone.[4] Communities along the Fitzroy and Burnett Rivers were particularly hard hit, while the Condamine, Ballone and Mary Rivers recorded substantial flooding. An unexpected flash flood raced through Toowoomba's central business district before devastating communities in the Lockyer Valley. A few days later thousands of houses in Ipswich and Brisbane were inundated as the Brisbane River rose and Wivenhoe Dam used a considerable proportion of its flood mitigation capacity. Volunteers were quick to offer assistance and sympathy was expressed from afar. A large mobilisation of the Australian Defence Force was activated and a relief fund created. The head of the recovery taskforce was Major General Michael Slater, DSC, AM, CSC. The Queensland Reconstruction Authority was formed to co-ordinate the rebuilding program beyond the initial taskforce and a Commission of Inquiry established to investigate all matters related to the floods.

The 2010–2011 floods killed 35 people in Queensland.[5] As of 26 January, an additional nine people were missing.[6] The state's coal industry was particularly hard hit. The Queensland floods were followed by the 2011 Victorian floods which saw more than fifty communities in western and central Victoria also grapple with significant flooding.

Background

Brisbane experienced major flooding in 1893 and 1974 as the Brisbane River broke its bank and inundated low lying areas. Towns including St George and Theodore had dealt with major flooding earlier in 2010.

The floods were a result of heavy rainfall caused by Tropical Cyclone Tasha that combined with a trough during the peak of a La Niña event. The 2010 La Niña weather pattern, which brings wetter conditions to eastern Australia, was the strongest since 1973. This La Niña event caused a prolonged event of heavy rainfall over Queensland river catchments. Record or near to record sea surface temperatures were recorded off the Queensland coast in late 2010. December 2010 was Queensland's wettest on record, with record high rainfall totals set in 107 locations for the month. The state average rainfall level of 209.45 millimetres (8.25 in) exceeded the previous record of 200.1 millimetres (7.88 in) set in 1975. 2010 was also recorded as the state's wettest spring since 1900 and the Australian continent's third wettest year. Kevin Trenberth thought climate change as a contributing factor in the unusually high precipitation rates. He attributed a half degree Celsius rise in ocean temperatures around Australia to global warming which produces extra water vapour and intensifies the monsoon. Other scientists say that it is too early to draw such a conclusion. Assertions were also made that mismanagement of the Wivenhoe Dam might be a contributor to the some flooding in the lower Brisbane valley although this is disputed by engineers.

Isolated flooding started across parts of the state in early December. On 23 December a monsoonal trough crossed the coast from the Coral Sea, bringing torrential rain that fell in a broad swath from the Gulf of Carpentaria to the Gold Coast. The widest range of intense rainfall occurred on 27 December with very high daily totals recorded on 25 December near where Cyclone Tasha crossed the coast By 30 December vast areas of Southern and Central Queensland were affected by the flood. The conditions led to a large influx of snakes in the Rockhampton area, as well as some crocodiles.

Extent

Flooding was widespread across Queensland from end of December 2010 to January 2011 with several separate rain events causing rivers to rise over a lengthy period.[7] Many places, including Condamine and Chinchilla were inundated by flood waters on multiple occasions.[8] About 300 roads were closed, including nine major highways.[9]

During the flooding it was reported that more than half of Queensland was affected by flooding.[8] [10]

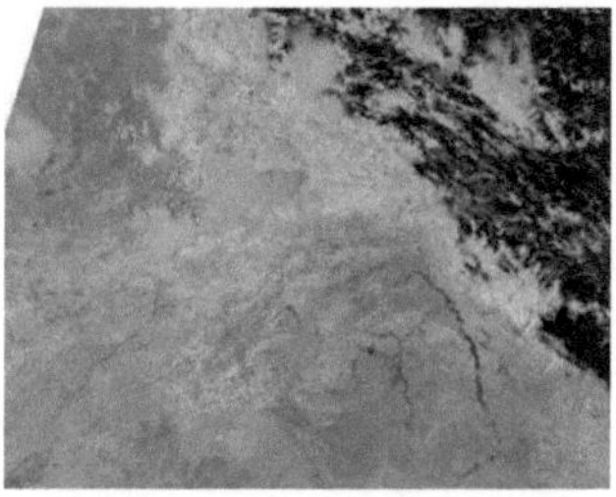

NASA image showing swollen rivers and cloud cover.

Fitzroy River basin

The flooding initially forced the evacuation of 1,000 people from Theodore and other towns, described as unprecedented by the acting chief officer of the Emergency Management Queensland.[11] The military transported residents by helicopter to an evacuation centre at Moura. The total evacuation of a Queensland town was a first for the state.[12] Major flooding at Theodore persisted for more than two weeks.[13]

Rockhampton seen from the air on 31 December; the Fitzroy River can be seen to have burst its banks.

Emerald was cut-off by road on 29 December as the Nogoa River rose.[14] By the next day, the river surpassed the 2008 flood peak level of 15.36 metres (50.4 ft).[15] At the peak of the flooding, 80% of the town was flooded, the worst the town ever experienced.[16] 1,200 Emerald residents registered as evacuees.[17]

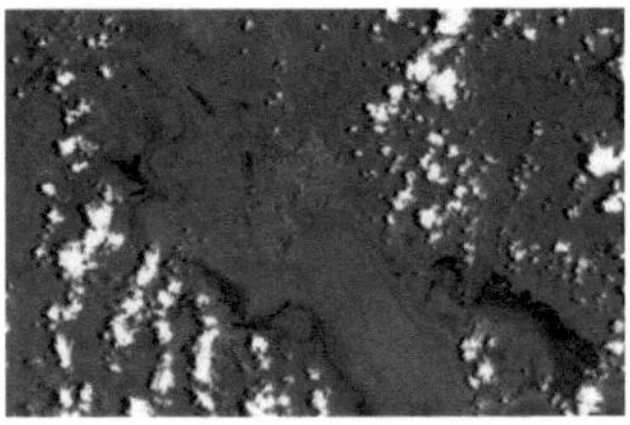
View of the swollen Fitzroy River, which surrounds the western half of Rockhampton.

Rockhampton had nearly a week to prepare for an expected flood peak from the Fitzroy River, which courses through the centre of the city.[18] The airport was closed on 1 January.[19] A metal flood barrier was erected around the terminal to prevent flood-borne debris from causing damage to the structure. An evacuation centre was set up at the Central Queensland University.[17] The Bruce Highway leading south out of Rockhampton was closed to traffic. The river peaked at 9.2 metres (30 ft) just short the of the predicted 9.4 metres (31 ft) maximum.[20]

The Port of Gladstone reduced its export capacity because the coal stockpiles at the port were saturated and further coal deliveries could not be made by rail.[21] The Goonyella railway line which services a number of coal mines in the Bowen Basin was closed for one week and shipments of grain were also delayed.[22]

Burnett River basin

The central Burnett towns of Gayndah and Mundubbera saw major flooding on 28–29 December. The Burnett River peaked at 18.25 metres (59.9 ft) at Mundubbera—the highest river height since 1942—inundating more than 20 houses. Downstream at Gayndah, the river peaked at 16.1 metres (53 ft) with floodwaters reaching two houses.[23] Both towns were isolated for several days and there was major disruption to the potable water supply and local agricultural production.[23] [24]

The swollen Burnett River at Gayndah, 350 kilometres (220 mi) north west of Brisbane.

Bundaberg experienced severe flooding, the worst in 40 years,[18] after the Burnett River flooded the city, although the Paradise Dam reservoir contributed to some flood mitigation.[7] This resulted in the evacuation of 300 homes.[18] The Bundaberg Port, a major sugar exporting facility, was closed late in December 2010 as flooding deposited silt in the port, forcing its closure. The port re-opened in early March 2011 after successful dredging operations allowed ships to berth.[25]

Condamine/Balonne River basin

Churchill Drive flooded in Warwick.

Chinchilla and Jericho were also inundated.[14] At least 40 residents were evacuated from Chinchilla.[26]

Flooding in Dalby was the worst since 1981.[27] The town's water purification system was flooded, resulting in water restrictions that have hampered clean-up efforts. 112500 litres (24700 imp gal; 29700 U.S. gal) of water were transported to the town of 14,000 residents.[28] Warwick was isolated when all roads into the town were cut-off.[29]

The Condamine River reached 14.25 metres (46.8 ft) on 30 December, its highest level ever recorded.[30] Condamine was mandatorily evacuated on 30 December.[31]

A second rain event on 9–10 January saw floodwaters again threaten Chinchilla and Condamine with Chinchilla residents again asked to evacuate.[32]

The Macintyre River, which forms part of the border between the Australian states of New South Wales and Queensland, began to threaten the town of Goondiwindi on 13 January. While the town is protected by a 11 metres (36 ft) levee, the local hospital and aged care home were evacuated as a precaution and an evacuation centre was established.[33]

Mary River basin

Heavy rain in the Mary River catchment on 8–9 January 2011 led to flooding at Maryborough and Gympie. The Mary River at Maryborough was expected to initially peak at 8.5 metres (28 ft) at midday 9 January with some houses and businesses inundated.

Toowoomba and the Lockyer Valley

Long and Mackenzie Streets in Toowoomba flooded.

The city of Toowoomba, in the Darling Downs, was hit by flash flooding after more than 160 millimetres (6.3 in) of rain fell in 36 hours to 10 January 2011; this event caused four deaths in a matter of hours.[32] [34] [35] Cars were washed away (see video [36]). Toowoomba sits on the watershed of the Great Dividing Range, some 700 metres (2300 ft) above sea level. A three week period where it had rained on all but three days had left the soil around Toowoomba super saturated and when a line of storms hit the city on 10 January, the resulting torrential rain rapidly ran off down gullies and streets. The central business district of the city sits in a small valley where two small water courses—East Creek and West Creek—meet to form Gowrie Creek. Unable to cope with the volume of water heading toward them, the creeks burst their banks, pushing a devastating wall of water through the city centre.[37] This water then headed west, not towards the Lockyer Valley which was also experiencing extreme rainfall that fell on eastern facing slopes.[38]

The surge passed through the Lockyer Valley town of Withcott, where the force of the water pushed cars into shops and forced the evacuation of hundreds of people. The scene was described by an onlooker as "like Cyclone Tracy has gone through it ... If you dropped an atom bomb on it, you couldn't tell the difference."[39] Nearby Helidon had several homes and farms flooded but did not break the main creek bank and enter the town. It was cutoff from all sides by destroyed roads. Grantham was also devastated by the surge of water. Houses were left crumpled by what

Premier of Queensland Anna Bligh described as an "inland tsunami". According to local media, the flood waters had reached a height of 7 or 8 m (23 or 26 ft) by the time it struck Grantham.[40] At least 100 people were evacuated to the Helidon Community Hall.[41] Nine people were confirmed dead, and many more feared dead among 66 reported missing.[42] The body of one victim washed away at Grantham was recovered 80 kilometres (50 mi) downstream and Queensland Police Commissioner Bob Atkinson warned that some bodies may never be found.[43] Nearby Gatton saw voluntary evacuations as the Lockyer Creek rose to a record height of 18.92 metres (62.1 ft), exceeding the previous record set in the 1893 Queensland floods.[32]

Brisbane River catchment

Flooding began to affect low-lying areas of Brisbane on the morning of 11 January 2011. By around 2:30 p.m. AEST, the Brisbane River broke its banks leading to evacuations in the Brisbane CBD and the suburbs of Fortitude Valley and West End.[44] [45] An evacuation centre was established for flood-affected residents at the RNA Showgrounds in Bowen Hills.[46] Residents of 2100 Brisbane streets were advised to evacuate prior to the arrival of floods, which struck the city on 12 January. Lord Mayor Campbell Newman stated than an estimated 20,000 homes would be affected when the river peaked on 14 January.[47] He subsequently advised that the Brisbane River transport infrastructure had been "substantially destroyed".[48]

Wheel of Brisbane during the floods.

The Brisbane River peaked on 13 January at a lower level than predicted,[8] [49] but still 20,000 houses in Brisbane were inundated. Some of the Brisbane suburbs worst affected by the floods were St Lucia, West End, Rocklea and Graceville.[50] The floods damaged some of Brisbane's icons. The Brisbane Riverwalk, a floating walkway over the Brisbane River linking the inner city neighbourhoods of Fortitude Valley and New Farm, broke up, with a section forming a 300 metres (980 ft) "floating missile" that threatened the Sir Leo Hielscher Bridges. The largest part of the floating boardwalk was safely guided under the bridge by a tugboat and past other infrastructure before being safely secured.[51] Brisbane's major Rugby league and Soccer venue, Suncorp Stadium, filled with water up to 2 metres (6.6 ft) deep, reaching the third or fourth row of seats.[52]

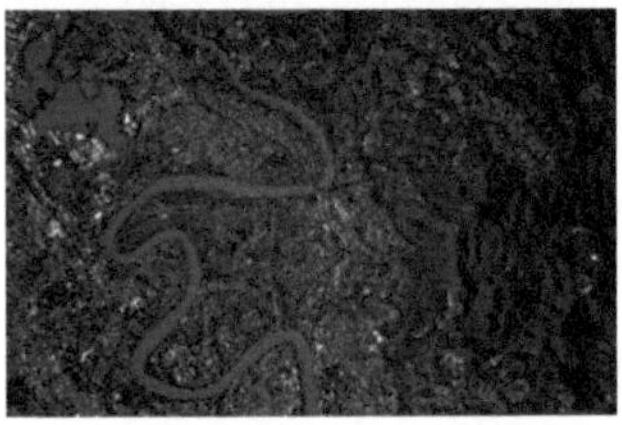

This detailed astronaut photograph illustrates flooding in suburbs of the Brisbane metropolitan region.

With the flood peaking at 4.46 metres (14.6 ft) in Brisbane City, the flood level was about the 10th highest in the city's history, several metres below the 1890 flood and the two major floods in 1893.,[49] [53] Parts of the western suburbs of Brisbane were cut off for three days. Resident of suburbs including Bellbowrie, Karana Downs, Moggill and Pullenvale were running low on food and other items when Moggill Road was cut, until the Australian Army was able to reach the area on 15 January with supplies.[54]

The Bremer River at Ipswich, 30 kilometres (19 mi) west of Brisbane, reached a height of 19.4 metres (64 ft) on 12 January, inundating the central business district and at least 3,000 houses.[55] One third of the city was reported to be underwater and over 1100 people took shelter at evacuation centres.[56] At nearby Marburg, a four year old boy was swept away by floodwaters when he fell from a rescue boat.[56] A man in his fifties died when he accidentally drove into floodwaters in the Ipswich suburb of Wulkuraka.[57] The worst affected areas of Ipswich were the suburbs of

Goodna and Gailes.[58] The flooding allowed bull sharks to reach the center of Goodna; one was spotted swimming in Williams Street, and a second in Queen Street.[59]

Elsewhere

In South East Queensland, the Wivenhoe Dam filled to a level equivalent to 191% of its supply capacity on 11 January 2011.[7] [60] Although the dam can hold the equivalent of 225% of its supply capacity, storage exceeding 100% is required to be progressively released through the floodgates to restore capacity to mitigate future flood events.[8] [61] [62] Brisbane experienced its wettest December since 1859.[21] Wyaralong Dam, near Beaudesert, had recently been completed and was praised[63] for mitigating flooding in downstream Logan, having exceeded 80% of its capacity.[64]

In North Queensland, the town of Ingham became isolated as the Herbert River peaked. Homes at Babinda, and Gordonvale were flooded.[65]

Deaths

As of 28 January 2011, 35 deaths have been attributed to the floods, 21 of which are from the Toowoomba and the Lockyer Valley area.[66] Additionally, 9 people are listed as missing, after the catastrophic flash flood struck Toowoomba, Lockyer Valley, Brisbane and Ipswich.[67] [68] [69] [70]

The first death was reported on 20 November 2010, when a woman drowned in her vehicle near Dysart.[71] In the following weeks, five others died from similar circumstances.[72] [73] [74] [75] [76] Nine people drowned after being caught directly by the flowing waters.[34] [77] [78] [79] [80] [81] and thirteen people died in other circumstances.[34] [57] [81] [82] [83] [84] [85] [86]

13 year old Jordan Lucas Rice died January 10, 2011 after insisting that his 10 year old brother Blake be saved from the flood waters before him. He drowned along with his mother, 43 year old Donna Maree Rice, both being residents of Toowoomba[87] He has been dubbed by the public as an Australian hero.[88] [89] [90]

Response

Public response

More than 55,000 volunteers registered to help clean up the streets of Brisbane,[91] with thousands more unregistered volunteers wandering the muddy streets with gumboots and mops.[92] Prime Minister Julia Gillard said the volunteering effort showed tremendous 'Aussie spirit' and that, "... right across Queensland today people have got up, they've marched out of their homes and they've gone to find people to help. It's a tremendous spirit of volunteering right across Queensland ...".[93]

Reaction

> The impact on our friends in Queensland is something that New Zealanders feel very deeply, and we just want to make sure that we're in a position to offer them support across the range of services that we have here in New Zealand.
>
> New Zealand Prime Minister John Key on the floods.[94]

Prime Minister of Australia Julia Gillard toured flood-affected areas on 31 December,[17] the day before The Queen (Elizabeth II), sent her expressions of concern and sympathy for the victims of the floods to her representative in Queensland, Governor Penelope Wensley.[95] [96] The Queen later made a donation to her Queensland premier's flood relief appeal,[97] as did her son, The Prince of Wales.[98] In March 2011, The Prince's eldest son, Prince William, toured the affected areas in Queensland, as well as those damaged by the floods in Victoria.[99]

US Secretary of State Hillary Clinton stated, "On behalf of President Obama and the people of the United States, I offer my condolences for the loss of life and damage in Queensland caused by the recent flooding."[100] U.S.

President Barack Obama said he was ready to help.[100]

John Key, the Prime Minister of New Zealand, held a telephone call with Gillard, during which she thanked New Zealand for its help. Len Brown, the Mayor of Auckland, a sister city of Brisbane, offered assistance to Brisbane Lord Mayor Campbell Newman.[101]

David Cameron, the Prime Minister of the United Kingdom offered his sympathies. Downing Street stated: "The Prime Minister offered his deep sympathy on behalf of the British people for those who have died, those who have lost their homes and all those who have been affected by the floods in Queensland."[102]

Recovery

> As we weep for what we have lost, and as we grieve for family and friends and we confront the challenge that is before us, I want us to remember who we are. We are Queenslanders. We're the people that they breed tough, north of the [New South Wales] border. We're the ones that they knock down, and we get up again. Together we can pull through this and that's what I'm determined to do, with your help, we can achieve it.
>
> Premier of Queensland Anna Bligh.[103]

Major General Michael Slater was appointed head of a civilian recovery task force.[20] Slater was at the time also the Commander of the 1st Division based at the Gallipoli Barracks in Enoggera, Brisbane. The taskforce was later replaced by the Queensland Reconstruction Authority, a new body created to co-ordinate the rebuilding program.[104]

The Australian Defence Force's contribution to flood relief efforts was designated Operation Queensland Flood Assist. The ADF established Joint Task Force 637, based at Enoggera Barracks in Brisbane, for operational command of on 1 January 2011.[105] According to the Australian Prime Minister the deployment was the largest for a natural disaster since Cyclone Tracy.[106]

About 35 State Emergency Service personnel from New South Wales and 20 personnel from Victoria were deployed to provide relief to exhausted staff and volunteers.[107] A national appeal was established on 29 December, with the state and federal governments giving A$1 million each.[30]

Small businesses and primary producers in 13 local government areas became eligible for grants of up to A$25,000 to pay for costs from damage incurred as a result of the floods.[108] National Disaster Relief and Recovery Arrangements were made available to a total of 31 local government areas across Queensland.[109]

New Zealand announced that the country would be sending two fifteen member civil defence teams to assist in flood rescue efforts.[110] In addition to this, New Zealand will send a number of New Zealand Defence Force soldiers, mainly engineers, as well as New Zealand Police and New Zealand Fire Service firefighters.[94] A New Zealand Red Cross team was also dispatched.[111]

Aid efforts

On 9 January, the *Flood Relief Appeal: Australia Unites* telethon broadcast by the Nine Network from the Suncorp Piazza at South Bank Parklands, Brisbane raised more than A$10 million in pledged aid.[112] The South Bank area which includes the Wheel of Brisbane pictured above was itself inundated by flood waters only two days later.

At the first Twenty20 cricket match between Australia and England in Adelaide on 12 January, both teams donated part of their match fees to help the victims and A$28,450 was collected from people in attendance.[113]

On 16 January, the day before the 2011 Australian Open commenced, a number of players took part in the *Rally for Relief* which was held at Rod Laver Arena and attended by 15,000 people, which raised A$750,000.[114] The Seven Network televised the event. Players who took part included Roger Federer, Rafael Nadal, Lleyton Hewitt, Novak Djokovic and Andy Roddick.[114]

On 27 January, the 2011 Legends of Origin charity match was contested between former New South Wales and Queensland Rugby League State of Origin players. Organised by Mark Geyer, the match raised $380,587 to go towards the recovery effort.[115]

Inquiry

On 17 January 2011, Anna Bligh announced a Commission of Inquiry into the 2010–11 Queensland floods.[116] The Commission's Terms of reference cover a wide array of related aspects and stipulate a final report is due in one year.[117] On August 1, 2011 the Commission handed an interim report containing 175 recommendations to Ms Bligh.[118]

Mandatory disaster insurance

In return for Senator Nick Xenophon's support of the Queensland Flood Levy, the Government will modify the Natural Disaster Relief and Recovery Arrangements and force the states to take up disaster insurance.[119]

Economic Impacts

Queensland Treasurer Andrew Fraser said it was not possible to put a figure on the damage; "other than to say the damage bill is going to start with a *b* and not an *m*".[120] [121] The floods will cost the Australian economy at least A$10 billion.[122] In figures released at the end of January 2011, the Insurance Council of Australia calculated that 38,460 individual claims were lodged with insurers which were worth A$1.51 billion.[123] Nearly half of those claims were for damage to homes and more than half were made by those living in Brisbane.

The price of food across Australia will go up.[124] The floods will cost supermarket chains tens of millions of dollars.[124] Some communities isolated by floodwaters experienced food shortages, and a rise in the cost of fruits and vegetables was reported shortly after.[31] Food supplies to northern Queensland were disrupted requiring groceries to be transported to Townsville by ship.[125]

Coal railway lines were closed and numerous mine sites flooded. According to an analysist at Macquarie Group almost all the available stockpiles of coking coal in Queensland were exhausted in late January 2011 due to reduced rail capacity.[126] About 15% of the state's annual output of coal production was lost and by late March 2011 recovery was progressing slowly.[127] Four months after the floods the Dalrymple Bay coal terminal was operating at half its capacity as the open-cut mines in the region continued to deal with de-watering issues.[128]

Swimming pool salt was in short supply as salt mines near Rockhampton were flooded.[129] Thousands of litres of milk had to be dumped because it couldn't be transported for processing.[130] The persistent wet conditions also caused health problems for livestock. There was severe damage to roadways in forestry plantations across wide areas of the south of the state.[131]

Prime Minister Julia Gillard announced that the government would impose a flood levy on people across Australia, to fund reconstruction works.[132]

The floods caused banana prices to soar across Australia. The price of bananas reached as high as $15 per kilogram.[133]

See also

- 1893 Brisbane flood
- 1974 Brisbane flood
- Floods in Australia
- List of disasters in Australia by death toll
- March 2010 Queensland floods

References

[1] "Flood costs tipped to top $30b" (http://www.abc.net.au/news/stories/2011/01/18/3115815.htm). *ABC News* (Australian Broadcasting Corporation). 18 January 2011. . Retrieved 18 January 2011.

[2] "Australia: Queensland floods spur more evacuations" (http://www.bbc.co.uk/news/world-asia-pacific-12097280). *bbc.co.uk* (BBC). 31 December 2010. . Retrieved 14 January 2011.

[3] "Floods force mass evacuations in Queensland, Australia" (http://www.bbc.co.uk/news/world-asia-pacific-12087870). *bbc.co.uk* (BBC). 29 December 2010. . Retrieved 14 January 2011.

[4] Hurst, Daniel (11 January 2011). "Three-quarters of Queensland a disaster zone" (http://www.brisbanetimes.com.au/environment/weather/threequarters-of-queensland-a-disaster-zone-20110111-19mf8.html). *Brisbane Times*. Fairfax Media. . Retrieved 14 January 2011.

[5] "Death toll from Queensland Floods" (http://www.police.qld.gov.au/News+and+Alerts/Media+Releases/2011/01/death_toll_jan24.htm). Queensland Police Service. 24 January 2011. . Retrieved 26 January 2011.

[6] Lisa Martin (26 January 2011). "Time for Queensland to heal after the floods – Queensland Premier Anna Bligh" (http://www.heraldsun.com.au/news/national/time-for-queensland-to-heal-bligh/story-e6frf7l6-1225994612338). *Herald Sun* (Herald and Weekly Times.). . Retrieved 26 January 2011.

[7] Hubert Chanson (2011). " The 2010–2011 Floods in Queensland (Australia): Observations, First Comments and Personal Experience (http://espace.library.uq.edu.au/view/UQ:239732)". *La Houille Blanche* (Paris: Societe Hydrotechnique de France) (1): 5–11. ISSN 0018-6368.

[8] Hubert Chanson (2011). "The 2010–2011 Floods in Queensland (Australia): Photographic Observations, Comments and Personal Experience" (http://espace.library.uq.edu.au/view/UQ:241445). *Hydraulic Model Report No. CH82/11* (Brisbane, Australia: The University of Queensland, School of Civil Engineering) (CH82/11): 127 pp. ISBN 9781742720234. .

[9] "Hundreds of roads closed in flooded Qld" (http://news.smh.com.au/breaking-news-national/hundreds-of-roads-closed-in-flooded-qld-20101228-198z7.html). *The Sydney Morning Herald* (Fairfax Media). 28 December 2010. . Retrieved 30 December 2010.

[10] 'La Niña and monsoonal winds flood northern Australia' (http://www.newscientist.com/article/mg20927941.400-la-nina-and-monsoonal-winds-flood-northern-australia.html), New Scientist, Jan. 8 2011, p. 6.

[11] Calligeros, Marissa (29 December 2010). "Flooding could last 'weeks, not days'" (http://www.brisbanetimes.com.au/environment/weather/flooding-could-last-weeks-not-days-20101229-199iy.html). *Brisbane Times* (Fairfax Media). . Retrieved 15 January 2011.

[12] "Flood destroys doctor's surgery in Theodore" (http://www.abc.net.au/rural/news/content/201101/s3109577.htm). *ABC Rural* (Australian Broadcasting Corporation). 10 January 2011. . Retrieved 15 January 2011.

[13] "Special Climate Statement 24" (http://www.bom.gov.au/climate/current/statements/scs24.pdf). Bureau of Meteorology. 7 January 2011. . Retrieved 18 January 2011.

[14] Calligeros, Marissa (30 December 2010). "From drought to flooding rains: Queensland's flood crisis continues" (http://www.brisbanetimes.com.au/environment/weather/from-drought-to-flooding-rains-queenslands-flood-crisis-continues-20101229-19a4p.html). *Brisbane Times* (Fairfax Media). . Retrieved 30 December 2010.

[15] "Qld's 'heartbreaking' floods worst ever" (http://news.smh.com.au/breaking-news-national/qlds-heartbreaking-floods-worst-ever-20101230-19adw.html). *The Sydney Morning Herald*. Australian Associated Press. 30 December 2010. . Retrieved 30 December 2010.

[16] Campion, Vikki (1 January 2011). "Flooded towns in Queensland declared disaster zones" (http://www.heraldsun.com.au/news/national/flooded-towns-in-queensland-declard-disaster-zones/story-e6frf7l6-1225979751206). *The Daily Telegraph* (News Limited). . Retrieved 1 January 2011.

[17] "Rockhampton faces forced evacuations" (http://www.abc.net.au/news/stories/2010/12/31/3104500.htm?section=justin). *ABC News Online* (Australian Broadcasting Corporation). 31 December 2010. . Retrieved 3 January 2011.

[18] Gorrie, Arthur (4 January 2011). "SES volunteers take goodwill gift" (http://www.gympietimes.com.au/story/2011/01/04/our-ses-volunteers-take-goodwill-gift/). *The Gympie Times* (APN News & Media). . Retrieved 4 January 2011.

[19] "Rockhampton Airport shuts down" (http://www.themorningbulletin.com.au/story/2011/01/03/rockhampton-airport-shuts-down/). *The Morning Bulletin* (APN News & Media). 3 January 2011. . Retrieved 3 January 2011.

[20] "Queensland floods: Rockhampton floodwaters 'peak'" (http://www.bbc.co.uk/news/world-asia-pacific-12116919). *bbc.co.uk* (BBC). 5 January 2011. . Retrieved 7 January 2011.

[21] Fickling, David; Brindal, Ray (30 December 2010). "Rains Hit Australian Coal Mines" (http://online.wsj.com/article/SB10001424052970203513204576048852232192150.html). *Wall Street Journal* (Dow Jones & Company). . Retrieved 30 December 2010.

[22] O'Sullivan, Matt (4 January 2011). "Losses and delays in flooded state" (http://www.smh.com.au/business/losses-and-delays-in-flooded-state-20110103-19dst.html). *The Sydney Morning Herald* (Fairfax Media). . Retrieved 4 January 2011.

[23] "Worst over for Gayndah, Mundubbera" (http://news.ninemsn.com.au/national/8189868/worst-over-for-gayndah-mundubbera). *NineMSN*. 29 December 2010. . Retrieved 4 January 2011.

[24] Koch, Tony (31 December 2010). "Disease next threat as barely a drop to drink in flooded Queensland towns" (http://www.theaustralian.com.au/news/nation/disease-next-threat-as-barely-a-drop-to-drink-in-flooded-queensland-towns/story-e6frg6nf-1225979003071). *The Australia* (News Limited). . Retrieved 4 January 2011.

[25] "Sugar backlog slowly clearing from Bundaberg port" (http://www.abc.net.au/rural/news/content/201103/s3171200.htm). *ABC Rural* (Australian Broadcasting Corporation). 23 March 2011. . Retrieved 4 June 2011.

[26] "Disasters declared in flood-hit Qld towns" (http://news.smh.com.au/breaking-news-national/disasters-declared-in-floodhit-qld-towns-20101228-198sn.html). *The Sydney Morning Herald*. Australian Associated Press. 28 December 2010. . Retrieved 4 January 2011.

[27] Trenwith, Courtney (30 December 2010). "Heartbreak beneath the flood slick" (http://www.brisbanetimes.com.au/environment/weather/heartbreak-beneath-the-flood-slick-20101229-19ad6.html). *Brisbane Times* (Fairfax Media). . Retrieved 30 December 2010.

[28] "Drinking water trucked into flood-ravaged Dalby" (http://www.heraldsun.com.au/news/breaking-news/drinking-water-trucked-into-flood-ravaged-dalby/story-e6frf7jx-1225978676970). *Australian Associated Press* (Herald Sun). 30 December 2010. . Retrieved 1 January 2011.

[29] O'Loan, James (28 December 2010). "Residents evacuate as flooding grips much of Queensland" (http://www.heraldsun.com.au/news/national/residents-evacuate-as-flooding-grips-much-of-queensland/story-e6frf7l6-1225977082293). *The Courier-Mail* (News Limited). . Retrieved 4 January 2011.

[30] "Disease fears grow as flooding hits towns' water supplies and rivers rise in Queensland" (http://www.news.com.au/national/disease-fears-grow-as-flooding-hits-towns-water-supplies-and-rivers-rise-in-rockhampton-bundaberg-and-emerald/story-e6frfkwr-1225978700048). *The Courier-Mail* (News Limited). 30 December 2010. . Retrieved 30 December 2010.

[31] "Some Queensland flood victims facing long wait to return to their flooded homes" (http://www.theaustralian.com.au/news/nation/some-queensland-flood-victims-facing-long-wait-to-return-to-their-flooded-homes/story-e6frg6nf-1225978512441). *The Australian* (News Limited). 30 December 2010. . Retrieved 30 December 2010.

[32] "Torrential rain raises alarm for Toowoomba, Maryborough, Gympie, Wide Bay, Kingaroy, Cooloola and Brisbane" (http://www.couriermail.com.au/news/torrential-rain-raises-alarm-for-maryborough-gympie-wide-bay-kingaroy-cooloola-and-brisbane/story-e6freon6-1225984664015). *Courier Mail* (News Limited). 10 January 2011. . Retrieved 10 January 2011.

[33] "Goondiwindi community flees ahead of rising river" (http://www.brisbanetimes.com.au/environment/weather/goondiwindi-community-flees-ahead-of-rising-river-20110113-19pvj.html). *Brisbane Times*. 13 January 2011. . Retrieved 13 January 2011.

[34] Atfield, Cameron; Hurst, Daniel (10 January 2011). "Four confirmed dead in Toowoomba flash flooding" (http://www.brisbanetimes.com.au/environment/weather/four-confirmed-dead-in-toowoomba-flash-flooding-20110110-19kti.html). *Australian Associated Press* (Brisbane Times). . Retrieved 10 January 2011.

[35] "Two dead as raging torrent swamps Toowoomba" (http://www.abc.net.au/news/stories/2011/01/10/3109846.htm). ABC News. 10 January 2011. . Retrieved 10 January 2011.

[36] http://www.theaustralian.com.au/news/breaking-news/desperate-and-grim-situation-facing-qld-premier-anna-bligh-warns/story-fn7il3q4-1225985275828

[37] Fraserq, Andrew (12 January 2011). "A once-in-a-century storm and an already saturated and hilly landscape were the ingredients for the tragedy that engulfed Toowoomba." (http://www.theaustralian.com.au/in-depth/queensland-floods/toowoomba-copped-wrath-of-la-nina/story-fn7iwx3v-1225985875512). *The Australian*. . Retrieved 12 January 2011.

[38] Graham Lloyd (15 January 2011). "An inland tsunami – like a fury from hell" (http://www.theaustralian.com.au/in-depth/queensland-floods/an-inland-tsunami-like-a-fury-from-hell/story-fn7iwx3v-1225988066591). *The Australian* (News Limited). . Retrieved 10 March 2011.

[39] Johnson, Callum (10 January 2011). "Withcott evacuated" (http://www.thechronicle.com.au/story/2011/01/11/withcott-flood-evacuation-toowoomba-rain/). *Toowoomba Chronicle*. . Retrieved 11 January 2011.

[40] "Grantham is a town left in tatters" (http://www.abc.net.au/news/video/2011/01/11/3110349.htm). ABC News. 11 January 2011. . Retrieved 14 January 2011.

[41] Amber Jamieson (11 January 2011), *'Situation continues to deteriorate': QLD floods* (http://www.crikey.com.au/2011/01/11/inland-instant-tsunami-devastates-southern-qld/),

[42] "Flood toll rises to 10, 78 missing" (http://www.couriermail.com.au/news/flood-toll-rises-to-nine-66-missing/story-e6freon6-1225985766915). *Australian Associated Press* (The Courier Mail). 11 January 2011. . Retrieved 11 January 2011.

[43] Ferguson, John; Schulz, Matthew (14 January 2011). "Queensland toll at 16 after body found" (http://www.heraldsun.com.au/ipad/queensland-toll-at-16-after-body-found/story-fn6bfm6w-1225987932899). *Herald Sun*. AAP and The Courier-Mail. . Retrieved 14 January 2011.

[44] "Raging floods bear down on Brisbane" (http://www.abc.net.au/news/stories/2011/01/11/3110450.htm). *Australian Broadcasting Corporation*. 11 January 2011. . Retrieved 11 January 2011.

[45] Hayden Donnell (11 January 2011). "Workers flee Brisbane city centre" (http://www.nzherald.co.nz/world/news/article.cfm?c_id=2&objectid=10699091). www.nzherald.co.nz. . Retrieved 11 January 2011.

[46] Dibben, Kay (11 January 20011). "Evacuation centre set up at Bowen Hills" (http://www.couriermail.com.au/ipad/evacuation-centre-set-up-at-bowen-hills/story-fn6ck51p-1225985695299). *The Courier-Mail* (News Limited). . Retrieved 11 January 20011.

[47] Long, Stephen (12 January 2011). "City braced for the worst: Brisbane Lord Mayor" (http://www.abc.net.au/pm/content/2011/s3111503.htm). *PM* (Australian Broadcasting Corporation). . Retrieved 13 January 2011.

[48] Malikin, Bonnie (13 January 2011). "Australian floods wreak havoc on Queensland economy" (http://www.telegraph.co.uk/news/worldnews/australiaandthepacific/australia/8256484/Australian-floods-wreak-havoc-on-Queensland-economy.html). *The Daily Telegraph* (London: Telegraph Media Group Limited). . Retrieved 14 January 2011.

[49] Richard Brown, Hubert Chanson, Dave McIntosh & Jay Madhani (2011). "Turbulent Velocity and Suspended Sediment Concentration Measurements in an Urban Environment of the Brisbane River Flood Plain at Gardens Point on 12–13 January 2011" (http://espace.library.uq.edu.au/view/UQ:243550). *Hydraulic Model Report No. CH83/11* (Brisbane, Australia: The University of Queensland, School of Civil Engineering) (CH83/11): 120 pp. ISBN 9781742720272. .

[50] "Police sent in as flood death toll rises" (http://www.abc.net.au/news/stories/2011/01/13/3112246.htm?section=justin). *ABC News*. Australian Broadcasting Corporation. . Retrieved 13 January 2011.

[51] Calligeros, Marissa (14 January 2011). "Riverwalk becomes 300m floating missile" (http://www.brisbanetimes.com.au/environment/weather/riverwalk-becomes-300m-floating-missile-20110112-19nwp.html). *Brisbane Times*. Australian Associated Press. . Retrieved 14 January 2011.

[52] McMahon, Bruce (14 January 2011). "Bubbling cauldron needs river to drop" (http://www.couriermail.com.au/ipad/bubbling-cauldron-needs-river-to-drop/story-fn6ck45n-1225987400161). *Courier Mail*. . Retrieved 14 January 2011.

[53] "Queensland Flood History" (http://www.bom.gov.au/hydro/flood/qld/fld_history/index.shtml). *Bureau of Meteorology*. . Retrieved 22 April 2011.

[54] Stolz, Greg; Chester, Darren (15 January 2011). "Help finally arrives in Brisbane's west" (http://www.couriermail.com.au/news/queensland/help-finally-arrives-in-brisbanes-west/story-fn7knuy7-1225988016944). *Courier Mail*. . Retrieved 15 January 2011.

[55] "Brisbane, Ipswich treading water" (http://www.adelaidenow.com.au/news/national/brisbane-ipswich-treading-water/story-e6frea8c-1225986072469). *Adelaide Advertiser*. 12 January 2011. . Retrieved 12 January 2011.

[56] "Bremer River at Ipswich peaks at 19.5m, lower than 1974 flood level" (http://www.couriermail.com.au/news/queensland/ipswich-residents-evacuate-ahead-of-bremer-river-peak/story-e6freoof-1225985598772). *Courier Mail*. 12 January 2011. . Retrieved 12 January 2011.

[57] Elks, Sarah (12 January 2011). "Ipswich body brings toll to 13" (http://www.theaustralian.com.au/in-depth/queensland-floods/ipswich-body-brings-toll-to-13/story-fn7iwx3v-1225986638717). *The Australian* (News Limited). . Retrieved 12 January 2011.

[58] "Queensland flood death toll rises to 12" (http://news.ninemsn.com.au/national/8195493/searchers-prepare-to-find-missing-in-qld). *NineMSN*. 12 January 2011. . Retrieved 12 January 2011.

[59] "Bull sharks seen in flooded streets" (http://www.qt.com.au/story/2011/01/14/ipswich-bull-sharks-spotted-flood-affected-streets/). *The Queensland Times*. 24 January 2011. . Retrieved 14 January 2011.

[60] "Controlled releases at Wivenhoe cut" (http://news.ninemsn.com.au/national/floods/8195758/controlled-releases-at-wivenhoe-cut). *Nine News* (NineMSN). 12 January 2011. . Retrieved 12 January 2011.

[61] "Wivenhoe Dam to release 12,000 cubic metres of water every second" (http://www.couriermail.com.au/news/wivenhoe-dam-gates-opened-wider-to-increase-flows-into-brisbane-river/story-e6freon6-1225985753811). *The Courier Mail*. 11 January 2011. . Retrieved 17 January 2010.

[62] "Wivenhoe and Somerset Dams" (http://www.cabinet.qld.gov.au/MMS/MediaAttachments/2010/pdf/32253 SEQWG Wivenhoe Fact Sheet A4 2pp F.pdf). Seqwater. . Retrieved 17 January 2010.

[63] Maizey, Judith (11 January 2011). "On the road in Logan today" (http://logan-west-leader.whereilive.com.au/news/story/on-the-road-in-logan-today/). *The Logan West Leader* (News Limited). . Retrieved 13 January 2011.

[64] Jackson, Zane (11 January 2011). "Wivenhoe Dam hits record high" (http://www.qt.com.au/story/2011/01/11/wivenhoe-dam-record-level-flood-rain-overflow/). *The Queensland Times* (APN News & Media Ltd). . Retrieved 13 January 2011.

[65] O'Loan, James; Robertson, Josh; MacDonald, Andrew; Michael, Peter; Higggins, Kate (28 December 2010). "Ten towns on flood alert as Queensland soaks up heavy rain" (http://www.couriermail.com.au/news/half-the-state-waterlogged-and-more-rain-to-come-with-expected-1b-damage-bill/story-e6freon6-1225976473249). *Australian Associated Press* (The Courier-Mail). . Retrieved 4 January 2011.

[66] "Death toll from Queensland floods" (http://www.police.qld.gov.au/News+and+Alerts/Media+Releases/2011/01/death_toll_jan24.htm). *Queensland Police* (Queensland Police). 24 January 2011. . Retrieved 14 January 2011.

[67] "Brisbane swamped as floodwaters rise" (http://www.abc.net.au/news/stories/2011/01/12/3111075.htm). *ABC News* (Australian Broadcasting Corporation). 12 January 2011. . Retrieved 12 January 2011.

[68] "Number of missing in floods revised down" (http://news.ninemsn.com.au/national/floods/8195697/number-of-missing-in-floods-revised-down). *Australian Associated Press* (NineMSN). 12 January 2011. . Retrieved 12 January 2011.

[69] "Flood death toll rises to 12 as killer tide continues its march" (http://www.brisbanetimes.com.au/environment/weather/flood-death-toll-rises-to-12-as-killer-tide-continues-its-march-20110112-19my3.html). *Brisbane Times* (Fairfax Media). 12 January 2011. . Retrieved 13 January 2011.

[70] "Fears for two families missing in floods" (http://www.abc.net.au/news/stories/2011/01/13/3111867.htm). *ABC News* (Australian Broadcasting Corporation). 13 January 2011. . Retrieved 13 January 2011.

[71] "Flood victim's body may have been found" (http://news.brisbanetimes.com.au/breaking-news-national/flood-victims-body-may-have-been-found-20101201-18gbm.html). *Australian Associated Press* (Fairfax Media). 1 December 2010. . Retrieved 9 January 2011.

[72] "Elderly man found dead in flood waters" (http://www.brisbanetimes.com.au/queensland/elderly-man-found-dead-in-flood-waters-20101205-18l08.html). *Australian Associated Press* (Brisbane Times). 5 December 2010. . Retrieved 9 January 2011.

[73] "Body found in car swept into flood waters" (http://news.brisbanetimes.com.au/breaking-news-national/body-found-in-car-swept-into-flood-waters-20101209-18qtk.html). *Australian Associated Press* (Brisbane Times). 9 December 2010. . Retrieved 9 January 2011.

[74] "Body found, towns prepare for more flood misery" (http://www.news.com.au/national/body-found-towns-prepare-for-more-flood-misery/story-e6frfkvr-1225980442362#ixzz1AW07yyPd). *news.com.au* (News Limited). 2 January 2011. . Retrieved 9 January 2011.

[75] "Man charged over Queensland flood death" (http://www.abc.net.au/news/stories/2011/01/07/3108325.htm). *ABC News* (Australian Broadcasting Corporation). 7 January 2011. . Retrieved 9 January 2011.

[76] Passmore, Daryl; Hinde, Suellen (9 January 2011). "Floods death count climbs to 11" (http://www.couriermail.com.au/ipad/floods-death-count-climbs-to-11/story-fn6ck51p-1225984299784). *The Sunday Mail (Qld)* (News Limited). . Retrieved 9 January 2011.

[77] Calligeros, Marissa (10 December 2010). "Body of drowned teen found" (http://www.brisbanetimes.com.au/queensland/body-of-drowned-teen-found-20101209-18r7i.html). *Brisbane Times* (Fairfax Media). . Retrieved 9 January 2011.

[78] "Body found in Brisbane River" (http://www.brisbanetimes.com.au/queensland/body-found-in-brisbane-river-20101223-196ds.html). *Australian Associated Press* (Brisbane Times). 20 December 2010. . Retrieved 9 January 2011.

[79] Calligeros, Marissa (30 December 2010). "Man found dead in flooded creek" (http://www.brisbanetimes.com.au/queensland/man-found-dead-in-flooded-creek-20101230-19aga.html). *Brisbane Times* (Fairfax Media). . Retrieved 9 January 2011.

[80] Kieran Campbell (12 January 2011). "Heartbreak over deaths in floods" (http://www.sunshinecoastdaily.com.au/story/2011/01/12/death-flood-family-children-water-rain-coast/). Sunshine Coast Daily. . Retrieved 12 January 2011.

[81] Ferrier, Tracey; Agius, Kym; Gray, Steve; Ja, Crystal (11 January 2011). "Qld flood death toll rises to 10" (http://news.smh.com.au/breaking-news-national/qld-flood-death-toll-rises-to-10-20110111-19lc5.html). *The Sydney Morning Herald*. Australian Associated Press. . Retrieved 11 January 2011.

[82] Calligeros, Marissa (13 December 2010). "Teenager drowned in arms of friends" (http://www.brisbanetimes.com.au/queensland/teenager-drowned-in-arms-of-friends-20101213-18uby.html). *Brisbane Times* (Fairfax Media). . Retrieved 9 January 2011.

[83] "Flood death as Rockhampton faces catastrophe with more cyclones predicted" (http://www.news.com.au/national/rockhampton-facing-catastrophe-as-more-cyclones-are-predicted/story-e6frfkvr-1225980725143#ixzz1AVyaYatS). *Australian Associated Press* (news.com.au). 3 January 2011. . Retrieved 9 January 2011.

[84] Maynard, Neale (11 January 2011). "Horrific scenes at Grantham" (http://www.news.com.au/breaking-news/floodrelief/horrific-scenes-at-grantham/story-fn7ik2te-1225985360479). *The Courier Mail* (news.com.au). . Retrieved 12 January 2011.

[85] "Devastated Toowoomba resembles 'ghost town'" (http://www.sheppnews.com.au/members/login.aspx?noaccess=1&from=/article.aspx?id=1196546). 11 January 2010. . Retrieved 11 January 2010.

[86] Colgan, Paul (11 January 2011). "24/7 Queensland floods info centre: updates, photos, advice and messages for loved ones" (http://www.news.com.au/breaking-news/floodrelief/flood-info-centre-updates-reports-warnings-advice-and-how-you-can-help/story-fn7ik2te-1225985436806). *news.com.au* (News Limited). . Retrieved 11 January 2011.

[87] "Queensland's flood heroes tell their story" (http://www.dailytelegraph.com.au/news/queenslands-flood-heroes-tell-their-story/story-e6freuy9-1225987437892). The Daily Telegraph. 14 January 2011. . Retrieved 18 January 2011.

[88] Travierso, Michelle (13 January 2011). "Heroism in the Floods: A Courageous Tale of Sacrifice From a Submerged Australia" (http://newsfeed.time.com/2011/01/13/the-sad-tale-of-the-young-hero-of-toowoomba/). *Time News Feed* (Time Inc). . Retrieved 18 January 2011.

[89] Box-Turnbull, Greig (17 January 2011). "Australia floods: Dad pays tribute to Jordan Rice, the teen who gave his life for his brother" (http://www.mirror.co.uk/news/top-stories/2011/01/17/australia-floods-dad-pays-tribute-to-jordan-rice-the-teen-who-gave-his-life-for-his-brother-115875-22854621/). *Daily Mirror* (Trinity Mirror). . Retrieved 18 January 2011.

[90] Reals, Tucker (13 January 2011). "Jordan Rice, 13, Tragic Hero of Australia Floods" (http://www.cbsnews.com/8301-503543_162-20028375-503543.html?tag=mncol;lst;1). *CBS News* (CBS Corporation). . Retrieved 18 January 2011.

[91] "Salvation comes with brooms and gumboots" (http://www.smh.com.au/environment/weather/salvation-comes-with-brooms-and-gumboots-20110115-19rv6.html). *The Sydney Morning Herald*. 16 January 2011. . Retrieved 18 January 2011.

[92] "Workers praised for tireless clean-up efforts" (http://www.abc.net.au/news/stories/2011/01/15/3113639.htm). *ABC News* (Australian Broadcasting Corporation). 18 January 2011. . Retrieved 18 January 2011.

[93] Transcript of press conference, Grafton 15 Jan 2011 - Prime Minister of Australia Press Office (http://www.pm.gov.au/press-office/transcript-press-conference-22) - retrieved 13 Dec 2011.

[94] TVNZ wire staff (13 January 2011). "Key: Economic impact of floods enormous" (http://tvnz.co.nz/politics-news/key-economic-impact-floods-enormous-3997059). *TVNZ* (Television New Zealand). . Retrieved 13 January 2011.

[95] "Queen offers sympathies to flood-hit Qld" (http://www.skynews.com.au/world/article.aspx?id=558917&vId=2063211). *Sky News* (Australian News Channel Pty Ltd). 1 January 2011. . Retrieved 3 January 2011.

[96] "Message following the flooding in Queensland, Australia, 31 December 2010" (http://www.royal.gov.uk/LatestNewsandDiary/Pressreleases/2010/MessagefollowingthefloodinginQueenslandAustralia31.aspx). *The official website of The British Monarchy* (The Royal Household). 31 December 2010. . Retrieved 4 January 2011.

[97] Associated Press (12 January 2011). "Queen makes private donation to aid Australian flood relief" (http://www.thestar.com/news/world/article/920866--queen-makes-private-donation-to-aid-australian-flood-relief). *Toronto Star*. . Retrieved 14 January 2011.

[98] Clarence House (14 January 2011). "News and Gallery > News > The Prince of Wales makes a donation to the flood relief efforts in Queensland" (http://www.princeofwales.gov.uk/newsandgallery/news/the_prince_of_wales_makes_a_donation_to_the_flood_relief_eff_93479592.html). Queen's Printer. . Retrieved 14 January 2011.

[99] "Prince William to visit Australia" (http://www.abc.net.au/news/stories/2011/03/09/3159032.htm). ABC. 8 March 2011. . Retrieved 8 March 2011.

[100] Townsend, Samantha; Rehn, Alison; Jones, Gemma (5 January 2011). "Obama ready to step in as Queensland flood crisis deepens" (http://www.dailytelegraph.com.au/news/obama-ready-to-step-in-as-queensland-flood-crisis-deepens/story-e6freuy9-1225981995822). *The Daily Telegraph*. News Limited. . Retrieved 5 January 2011.

[101] TVNZ wire staff (12 January 2011). "Auckland offers support to flood-ravaged sister city" (http://tvnz.co.nz/national-news/auckland-offers-support-flood-ravaged-sister-city-3997019). *TVNZ* (Television New Zealand). . Retrieved 13 January 2011.

[102] "Boy, 13, sacrifices his life to save brother in raging Australian floods" (http://www.dailymail.co.uk/news/article-1346418/Australia-floods-Queenslands-Jordan-Rice-13-sacrifices-life-save-brother.html). London: Daily Mail. 2011-01-12. . Retrieved 2011-01-26.

[103] "Flood death toll rises as major search gets under way for the missing" (http://www.adelaidenow.com.au/ipad/brisbane-flood-alert-as-wivenhoe-threatens-to-spill-over/story-fn6t2xlc-1225985477271). *Adelaide Advertiser* (News Limited). 13 January 2011. . Retrieved 13 January 2011.

[104] Jessica Marszalek (19 January 2011). "New body set up to rebuild Queensland" (http://www.brisbanetimes.com.au/queensland/new-body-set-up-to-rebuild-queensland-20110119-19w23.html). *Brisbane Times* (Fairfax Media). . Retrieved 31 January 2011.

[105] "Operation QUEENSLAND FLOOD ASSIST" (http://www.defence.gov.au/opEx/global/opqldflood/index.htm). Department of Defence. .

[106] "Gillard announces troop boost for floods" (http://news.smh.com.au/breaking-news-national/gillard-announces-troop-boost-for-floods-20110114-19q77.html). *Sydney Morning Herald* (Fairfax Media). 14 January 2011. . Retrieved 17 January 2011.

[107] Bligh, Anna (31 December 2010). "Other states sending aid to Queensland" (http://www.cabinet.qld.gov.au/MMS/StatementDisplaySingle.aspx?id=73156). *Ministerial Media Statement*. Queensland Government. . Retrieved 5 January 2011.

[108] "Financial assistance for primary producers" (http://www.business.qld.gov.au/risk-management/flood-farm-financial.html). Queensland Government. . Retrieved 7 January 2011.

[109] Roberts, Neil (27 December 2010). "Disaster Relief and Recovery Arrangements extended" (http://www.cabinet.qld.gov.au/MMS/StatementDisplaySingle.aspx?id=73131). *Ministerial Media Statement*. Queensland Government. . Retrieved 5 January 2011.

[110] TVNZ wire staff (13 January 2011). "Second NZ civil defence team headed to Qld" (http://tvnz.co.nz/national-news/more-nz-help-queensland-3997371). *TVNZ* (Television New Zealand). . Retrieved 13 January 2011.

[111] TVNZ wire staff (12 January 2011). "NZ red cross team named for Oz" (http://tvnz.co.nz/national-news/nz-red-cross-team-named-head-australia-3996497). *TVNZ* (Television New Zealand). . Retrieved 13 January 2011.

[112] "Almost $30m raised in Qld flood appeal" (http://www.abc.net.au/news/stories/2011/01/09/3109226.htm). *ABC News* (Australian Broadcasting Corporation). 10 January 2011. . Retrieved 10 January 2011.

[113] Summerford, Matt (13 January 2011). "Match fees donation will help victims of Queensland floods" (http://www.independent.co.uk/sport/cricket/match-fees-donation-will-help-victims-of-queensland-floods-2183014.html#). *The Independent* (London). . Retrieved 13 January 2011.

[114] Quayle, Emma (17 January 2011). "Open hearts: rivals rally around to help raise funds for flood relief" (http://www.smh.com.au/sport/tennis/open-hearts-rivals-rally-around-to-help-raise-funds-for-flood-relief-20110116-19skb.html). *Sydney Morning Herald* (Fairfax Media). . Retrieved 17 January 2011.

[115] Legends Of Origin Match Home | NRL | Triple M (http://www.triplem.com.au/sydney/sport/nrl/legends-of-origin-2011)

[116] "Flood inquiry announced as death toll rises" (http://www.abc.net.au/news/stories/2011/01/17/3114735.htm). *ABC News Online* (Australian Broadcasting Corporation). 17 January 2011. . Retrieved 17 January 2011.

[117] "Queensland's flood inquiry: the terms of reference" (http://www.brisbanetimes.com.au/environment/weather/queenslands-flood-inquiry-the-terms-of-reference-20110117-19tv7.html). *Brisbane Times* (Fairfax Media). 17 January 2011. . Retrieved 17 January 2011.

[118] Jodie van de Wetering (1 August 2011). "175 recommendations in interim flood report" (http://www.webcitation.org/60dLg0Ocs). *Australian Broadcasting Corporation* (Australian Broadcasting Corporation). . Retrieved 2 August 2011.

[119] Wilson, Lauren (23 March 2011). "Gillard's flood and cyclone levy wins vote" (http://www.theaustralian.com.au/national-affairs/gillards-flood-and-cyclone-levy-wins-vote/story-fn59niix-1226026374142). .

[120] Rogow, Geoffrey (11 January 2011). "Australian Dollar Down Late As Flood Impact Increases" (http://online.wsj.com/article/BT-CO-20110111-700158.html). *The Wall Street Journal*. Dow Jones Newswires. . Retrieved 41 January 2011.

[121] Lee, Jane; AAP (12 January 2011). "Queensland floods 'could wipe $13 billion off economic growth'" (http://www.news.com.au/business/queensland-floods-could-cut-gdp-by-1-per-cent/story-e6frfm1i-1225986231656l). *news.com.au* (News Limited). . Retrieved 14 January 2011.
[122] "Flood costs tipped to top $30b" (http://www.abc.net.au/news/stories/2011/01/18/3115815.htm). *ABC News Online* (Australian Broadcasting Corporation). 18 January 2011. . Retrieved 25 January 2011.
[123] Geoff Easdown (31 January 2011). "Queensland flood insurance claims top $1.5b" (http://www.heraldsun.com.au/business/queensland-flood-insurance-claims-top-15b/story-e6frfh4f-1225997292454). *Herald Sun* (Herald and Weekly Times). . Retrieved 6 February 2011.
[124] Michael Evans (25 January 2011). "Woolies slims a bullish forecast down to 5-8%" (http://www.theage.com.au/business/woolies-slims-a-bullish-forecast-down-to-58-20110124-1a2z9.html). *The Age* (Melbourne: Fairfax Media). . Retrieved 25 January 2011.
[125] Trenwith, Courtney (4 January 2011). "Hercules to the rescue" (http://www.brisbanetimes.com.au/environment/weather/hercules-to-the-rescue-20110104-19ewc.html). *Brisbane Times* (Fairfax Media). . Retrieved 4 January 2011.
[126] Matt Chambers (24 January 2011). "Coking coal prices to skyrocket as stockpiles dry up" (http://www.theaustralian.com.au/business/coking-coal-prices-to-skyrocket-as-stockpiles-dry-up/story-e6frg8zx-1225993253543). *The Australian* (News Corporation). . Retrieved 25 January 2011.
[127] "Floods cut Australia's Queensland coal output 30 mln/t-council" (http://af.reuters.com/article/metalsNews/idAFL3E7EV48220110401). *Reuters Africa* (Thomson Reuters). 1 April 2011. . Retrieved 1 April 2011.
[128] "Queensland's Dalrymple Bay coal terminal runs at only 50% of capacity" (http://www.platts.com/RSSFeedDetailedNews/RSSFeed/Coal/8839228). *Platts*. 3 May 2011. . Retrieved 5 May 2011.
[129] Jessica Johnston (4 January 2011). "Coast salt supplies dry up after floods" (http://www.goldcoast.com.au/article/2011/01/04/280095_gold-coast-news.html). *goldcoast.com.au* (News Limited). . Retrieved 25 January 2011.
[130] Brad Markham (28 January 2011). "Floods hit farmers in the pocket" (http://www.abc.net.au/local/stories/2011/01/28/3123709.htm). *ABC Wide Bay* (Australian Broadcasting Corporation). . Retrieved 31 January 2011.
[131] Jo Skinner (24 January 2011). "Drivers warned to avoid flood-hit plantation roads" (http://www.abc.net.au/news/stories/2011/01/24/3119613.htm). *ABC News Online* (Australian Broadcasting Corporation). . Retrieved 31 January 2011.
[132] Levy, Megan (27 January 2011). "Levy to pay for $5.6b flood bill" (http://www.brisbanetimes.com.au/business/levy-to-pay-for-56b-flood-bill-20110127-1a64x.html). *Brisbane Times*. . Retrieved 27 January 2011.
[133] http://www.abc.net.au/news/2011-05-27/high-banana-prices-may-stay-expert-says/2733200

External links

- Flooding in Australia (http://earthobservatory.nasa.gov/NaturalHazards/event.php?id=48320) and Flooding in Rockhampton, Queensland (http://earthobservatory.nasa.gov/IOTD/view.php?id=48456) at NASA Earth Observatory
- Flash flood death toll rises to 9 (http://www.abc.net.au/news/stories/2011/01/11/3110667.htm)
- Brisbane City Council COP Situational Awareness Application (http://elbflood-473794511.ap-southeast-1.elb.amazonaws.com/floodcop/), interactive map showing extent of flooding in Brisbane
- Interactive map of pre and post flood aerial images along the Brisbane River (http://www.nearmap.com/?ll=-27.508638,152.974365&z=21&t=k&nmd=20110113)
- Dam operator blames inaccurate weather forecast for 2011 Brisbane flood (http://www.abc.net.au/news/stories/2011/03/08/3157741.htm)
- Interim report of Queensland Floods Commission of Enquiry, handed to Premier Bligh on August 1 2011 (http://www.webcitation.org/60dMBtuqg)

Ipswich,_Queensland

Ipswich Ipswich, Queensland	

Ipswich CBD & Bremer River as seen from Riverlink Shopping Centre, North Ipswich

Population:	168,131(2010)[1]
Established:	1846
Postcode:	4305
Elevation:	50 m (164 ft)
Time zone:	AEST (UTC+10)
Location:	40 km (25 mi) from Brisbane CBD
LGA:	City of Ipswich
Region:	South East Queensland
State District:	Ipswich, Ipswich West, Bundamba, Part of Lockyer
Federal Division:	Blair, Oxley, Wright

Suburbs around Ipswich:		
Woodend	North Ipswich	East Ipswich
West Ipswich	**Ipswich**	Tivoli
One Mile	Churchill	Raceview

Ipswich is a city in South-East Queensland, Australia. Along the Bremer River Valley, it is approximately 40 km away from the state's capital, Brisbane. The suburb by the same name forms the city's Central Business District and administrative centre. Ipswich has a population of 163,383.

Urban sprawl from the mid-20th century onwards has made it become part of the Greater Brisbane conurbation. However, it has retained municipal status and has an independent history and culture. Ipswich is a major commercial and industrial area that is currently undergoing major transit oriented urban renewal, which was first planned in the Ipswich Regional Centre Strategy.[2] Ipswich was proclaimed as a municipality on 2 March 1860, and became a city in 1904.

History

Prior to the arrival of European settlers, what is now called Ipswich was home to multiple tribal lands, including the Yuggera and Ugarapul Indigenous Australian groups.[3] The area was first explored by European colonists in 1826, when Captain Patrick Logan, Commandant of the Moreton Bay penal colony, sailed up the Brisbane River and discovered large deposits of limestone and other minerals.[4]

Ipswich Railway Station in 1865

In 1826, the town began as a limestone mining settlement and grew rapidly as a major inland port, Ipswich was at first named Limestone. In 1843 it was renamed after the town of Ipswich in England. It became a municipality in 1858. Originally Ipswich was going to be the capital of Queensland, but in 1859, it made a failed bid, losing to rival Brisbane.[5] It was proclaimed a city in 1904.[6]

The Bremer River and Ipswich in 1872

In 1827, the first convicts and overseers arrived to mine the area. The town that built around this industry became known as Limestone and is remembered by one of the main streets through the Ipswich city centre.[7] In 1843, the settlement officially became known as Ipswich, making it the oldest provincial city in Queensland. The name change was proposed by surveyor Henry Wade and approved by Sir George Gipps, a soldier and Governor of New South Wales.[8] The town of Ipswich in England was once called Gipeswic,[9] derived from the Old English for "workplace by water".

Brisbane Street at the turn of the century

From the 1840s onward, Ipswich was becoming an important river port for the growing local industries such as coal, and wool from the Darling Downs, so a regular paddlesteamer service from Brisbane Town, 'The Experiment', was established in 1846.[10] This, and other steamer services,[11] remained the primary form of mass/bulk transport between the two cities until 1876, when the construction of the original Albert Bridge, spanning the Brisbane River at Indooroopilly, completed the railway line begun between Ipswich and Brisbane in 1873.[12]

Father William McGinty, a native of Northern Ireland, took charge of Ipswich Roman Catholic Parish in 1852 and had the first St Mary's Church built. This building was described as "a rude slab building of small dimensions, without glass windows, only shutters." This church was later demolished and replaced by a "beautiful stone edifice of Gothic design" as quoted from a member of the church community. That was the old St. Mary's and it cost £7,000 in 1862.

Ipswich was proclaimed as a municipality on 2 March 1860 and became a city in 1904. Damaging flooding has occurred on numerous occasions, the largest being the 1893 Brisbane flood peaking at 24.5 m, and more recently during the 1974 Brisbane Flood, peaking in Ipswich at 20.7 m. Collectively, 8500 houses were flooded in Ipswich and Brisbane.[13]

In the 1980s and the 1990s, with land in nearby Brisbane becoming fully developed, and very expensive for housing, many new large scale housing estates became attractive for buyers in Ipswich with lower prices and bigger lots. Such new housing estates sprung up in Goodna, Brassall, Springfield Lakes, Collingwood Park, Redbank Plains with acreage estates in Karalee, Barellan Point, and Chuwar. The first masterplanned community in South-east Queensland was subsequently established at Springfield.

Climate

Ipswich experiences a humid subtropical climate (Köppen climate classification *Cfa*).

Economy

Ipswich was a major mining centre, particularly coal mining. The city is the 'cradle of coal mining in Queensland'.[14] Other secondary manufacturing industries included earthenware works, sawmills, abattoirs and foundries, while the region is also rich agriculturally.

St Mary's Catholic church, Ipswich, Queensland.

Ipswich is the site of RAAF Base Amberley, the Royal Australian Air Force's largest operational base. RAAF Amberley is currently home to No. 1 and No. 6 Squadrons flying the Boeing F/A-18F Super Hornet. No. 36 Squadron (flying C-17 Globemaster III transport aircraft) are also based out of Amberley. In addition a number of ground support units are located at Amberley. The airbase is currently being expanded and will eventually also house No. 33 Squadron (flying Airbus A330 MRTT tanker-transports) and several Army units.

Ipswich is also home to a campus of the University of Queensland, the annual Ipswich Cup horse races one of the biggest race meets in Queensland where over 24,000 attend, and The Workshops Railway Museum [15] to commemorate the first trainline in Queensland from Ipswich to Grandchester (Bigges Camp), approximately 25 km to the west. The present Ipswich Station is a major transport interchange.

A large shopping complex is being developed in the eastern suburb of Springfield. Accompanying this is a satellite campus of the University of Southern Queensland, which has its primary campus in Toowoomba. Redbank Plaza shopping mall has had eight cinemas added as well as a fourth floor of shopping. Three clear retail and service hubs are developing outside Ipswich central and besides Springfield. Brassall and Karalee to the north, Yamanto to the South, and Booval in the central east.

Housing

The traditional Ipswich dwelling has always been a detached home on land, however this is changing as housing demand and a desire to live closer to amenities increases. New estates and suburbs are currently under development in the south-eastern section of the city as well as Ripley Valley.

Queensland style architecture, Ipswich, QLD

Infrastructure

Health

Ipswich Hospital is the major public hospital in the city. St Andrew's Private Hospital is the city's only major private hospital.

Transport

The main form of transport in Ipswich is the private vehicle. Ipswich also has direct access to the Ipswich Motorway (linking to Brisbane); the Cunningham Highway (linking to Warwick); the Warrego Highway (linking to Toowoomba); and the Centenary Highway (linking Springfield and the Ripley Valley to Brisbane).

Ipswich Railway Station is a major hub for rail transport. The electified rail lines that runs from Ipswich through Brisbane's western suburbs to the Brisbane CBD is known as the Ipswich Line. The Rosewood railway line is also electrified and continues west through Ipswich's western suburbs to the town of Rosewood. Both lines are operated by Queensland Rail. A rail line has been proposed to run from Ipswich to Yamanto, and the Ripley Valley. The Westlander, operated by Traveltrain stops at Ipswich Station twice a week on its journey to and from Charleville and stops along the way include Toowoomba.

Ipswich Railway Station

Ipswich is also the hub of a large bus network. Westside Buslines is the prominent bus company in Ipswich, linking all sides of town to Ipswich CBD; Redbank; Springfield; or Forest Lake (in Brisbane). They also operate services on behalf of Queensland Rail to Fernvale, Lowood, Coominya, Esk, and Toogoolawah. Southern Cross Citilink is a bus that operates between Ipswich and Indooroopilly via Riverlink, Karalee, Chuwar, Karana Downs, Anstead, and Kenmore. Bus Queensland Lockyer Valley operates bus services on behalf of Queensland Rail to Gatton and Helidon. An Airport Express Bus operates from Ipswich to Brisbane Airport on a regular basis. Prior bookings are essential though. Greyhound Australia offers daily direct services to Toowoomba, Mount Isa, Charleville, Sydney and Melbourne from the Ipswich Transit Centre. Crisps coaches offer daily direct services to and from Warwick and Stanthorpe.

Bicycle transport is not well facilitated in Ipswich, although there are some marked on-road lanes there are few segregated bicycle facilities. A 65 km trail linking Ipswich to Boonah includes bicycle and walking paths is currently under construction.[16] A new cycleway has been created at Brassall. The bikeway is 2.6 km long and connects the suburbs of Brassall and North Ipswich. Further bikeways are being constructed, such as the link to Redbank Plains High school and by the utilization of the former Brisbane Valley Rail line , which will connect Wulkuraka to Blackbutt.

Utilities

Water for Ipswich is supplied by Queensland Urban Utilities and is sourced from the South East Queensland Water Grid. Ipswich is under permanent water restrictions.

Sport

- The Ipswich Knights play in the Brisbane Premier League Division 1 competition.
- The Ipswich Jets play in the Queensland Cup competition.
- The V8 Supercars race each year at Queensland Raceway at Willowbank in mid July.
- Brookwater Golf Club, designed by Greg Norman, is located in Springfield.
- The Ipswich Cougars play in the Gridiron Queensland league and have been one of the most dominant teams since the launch of Gridiron in Queensland.
- AFL is well represented. The Ipswich Eagles play in second tier division of AFL Queensland, and Collingwood Park Power play in Division 3. The Ipswich Miners play in the under 18 competition of AFL Queensland. Collingwood Park(Power), Ipswich Central (Cats)and Springfield Lakes play in the junior competition of AFL Brisbane Juniors.

- Drag Racing is conducted at Willowbank Raceway year round with events open to both professional and amateur racers, including a series conducted to facilitate newcomers to the sport.
- Hockey is a major sport in the city and is played at Briggs road facility where there are 2 artificial surfaces that are world class.

Notable Residents

Sportspeople

- Allan Langer (attended Ipswich State High School and Blair State Primary School)
- The "Walters Brothers", rugby league players: Kevin Walters, Kerrod Walters and Steve Walters
- Dale Shearer, rugby league player who represented Queensland and Australia at rugby league
- Craig McDermott and Shane Watson, represented Queensland and Australia in cricket, attended Ipswich Grammar School.
- John Buchanan, former Australian cricket team coach
- Berrick Barnes, dual code rugby footballer and Wallabies player (attended Ipswich Grammar School)
- Rhan Hooper, professional Australian rules footballer in the Australian Football League (AFL) (attended Bundamba State Secondary College)
- Mark Tonelli, Olympic gold medallist in swimming, member of the *Quietly Confident Quartet.*
- Israel Folau, professional dual code footballer Australian rugby league representative and Australian rules footballer with the Greater Western Sydney Football Club. Played for Ipswich junior rugby league team Goodna Eagles.[17]
- Lagi Setu: NRL Player for The Brisbane Broncos (attended Ipswich Grammar School)
- Des Morris, Queensland rep and current Australian selector (attended Bremer High School)
- Rod Morris, former Balmain, Queensland, NSW and Australian Front
- Lama Tasi, NRL Player for The Sydney Roosters (attended Ipswich Grammar School)
- Luke Poulton: Futsal Player for The Australian National Futsal team (attended St Peter Claver College)

Politicians

- Neville Bonner: The first indigenous Australian to become a member of the Parliament of Australia.
- Llewellyn Edwards: Sir Llew entered Queensland state parliament as the Liberal Party Member for Ipswich Liberal Party of Australia in 1972. In 1984 he was made a Knight Bachelor, and in 1989, a Companion of the Order of Australia.
- Bill Hayden: former Governor General of the Commonwealth of Australia, leader of federal Australian Labor Party.
- Pauline Hanson: former member of the Liberal Party of Australia, Independent member for Oxley (1996–1998) and founder of the One Nation Party.
- Cameron Thompson: former Liberal member for Blair (1998–2007).
- Shayne Neumann: present Labor member for Blair (2007–present)
- David Hamill, attended Bremer High School and was dux. He was a Queensland labour minister in the Goss government

Musicians

- Matthew Hickey: singer, founder and director of The Ten Tenors
- David McCormack: former lead member of Custard and The Titanics.
- Harold Blair: Aboriginal Tenor and music teacher
- Thomas Shapcott: Author and poet
- Kharizma: contestants on the second season of *The X Factor.*

Other

- John Birmingham: (Liverpool-born) Author
- John Bradfield: Engineer and designer of the Sydney Harbour Bridge; Story Bridge; and the Sydney Railway System
- Joy Chambers: Author, Actor
- D'Arcy Doyle: Painter
- Sir Harry Gibbs: Chief Justice of the High Court of Australia
- Sir Samuel Griffith: Chief Justice of the High Court of Australia
- Frederick Kesner: (Manila-born) Poet
- Ben Atkins: General Legend

Education

Colleges and universities

- University of Queensland, Ipswich
- Auto Trade College, Bundamba
- Bremer Institute of TAFE, Ipswich, Bundamba and Goodna
- University of Southern Queensland, Springfield

Secondary schools

Ipswich has a considerable number of secondary schools, including the following:

- Bremer State High School [18]
- Bundamba State Secondary College
- Ipswich Girls' Grammar School, established 1892
- Ipswich Grammar School, established in 1863, first school to open as a result of the Grammar Schools Act of 1860
- Ipswich State High School
- Redbank Plains State High School
- Rosewood State High School [19]
- Saint Edmund's College [20]
- Saint Mary's College
- St Augustine's College Springfield
- St Peter Claver College [21]
- West Moreton Anglican College, Karrabin
- Westside Christian College [22]
- Woodcrest State College [23]

Primary schools

Ipswich has a number of primary schools, including the following:

- Amberley District State Primary School (replacing Amberley State Primary School)
- Goodna State School, established in 1870.
- Ipswich State School
- Sacred Heart Primary School
- Silkstone State School
- St Marys Primary school
- St Josephs Primary School
- Brassall State Primary School
- North Ipswich State Primary School
- East Ipswich State Primary School
- West Ipswich State Primary School
- Central State Primary School
- Bundamba State Primary School
- Blair State Primary School
- Karalee State Primary School
- Tivoli State Primary School
- Churchill State Primary School
- Redbank Plains State School
- Redbank State Primary School
- Riverview State School
- Camira State School

Events

Ipswich Show

The first Ipswich Annual Show was held on 2 April 1873 by the Queensland Pastoral and Agricultural Society. There had been shows staged by the Ipswich and West Moreton Horticultural and Agricultural Society as early as 1868. Originally held at the sale yards situated at Lobb St, Churchill, the show moved to its present home at the Ipswich Showgrounds in 1877.

Goodna Jacaranda Festival

This festival has been held annually at Goodna (Evan Marginson Sportsground) since 1968, over 10,000 people attend this three day event. The festival features a talent quest, school band competition, fun run, live entertainment as well as the regular attractions such as sideshow alley, food and stalls.

Since 2009 the festival has had its own website.[[24]. In 2010, it joined Facebook.[[25]

Goodna Jacaranda Festival Sign

Ipswich Festival

The Ipswich Festival is an annual multi-disciplinary festival that fosters and encourages an atmosphere, which brings families together to enjoy the many free attractions. The Ipswich Festival showcases a broad and bold program of entertainment and arts that continues to entertain, inspire and celebrate the essence of our City's culture, spirit and community to the highest calibre.

First staged in 1998, the Ipswich Festival has enjoyed a 13-year development strategy, and the planning, programming and production are professionally managed by an experienced team of individuals who are highly regarded as specialists in creating and managing festivals and community event.

The Ipswich Festival is regarded as the City's premier event and over the past 13 years, the variety, size and scope of the program has grown exponentially through increasing interest group participation, corporate partnerships and community enthusiasm through the presentation of a festival that `celebrates its city'.

The festival presents Ipswich's cultural heritage and highlights the city's emerging talents through a deliberate program of Live Bands, Concerts, Art exhibits, Fireworks, Theatre, Jazz, Multicultural Celebrations, interactive displays and a range of family-based events to encourage participation by all age groups whilst attracting day tourism to the region.

The Ipswich Festival runs for two weeks at the end of April beginning of May with the majority of events free.[26]

References

[1] Australian Bureau of Statistics (25 October 2007). "Ipswich (Ipswich City) (State Suburb)" (http://www.censusdata.abs.gov.au/ABSNavigation/prenav/LocationSearch?collection=Census&period=2006&areacode=SSC31279&producttype=QuickStats&breadcrumb=PL&action=401). *2006 Census QuickStats*. . Retrieved 2009-04-21.

[2] http://www.dip.qld.gov.au/local-area-planning/transit-oriented-development.html

[3] http://www.ipswich150.com.au/proud_past

[4] Extract from Captain Patrick Logans' Journal - The Sydney Gazette and New South Wales Advertiser (NSW : 1803-1842) (http://nla.gov.au/nla.news-article2188791)

[5] http://www.epa.qld.gov.au/chims/placeDetail.html?siteId=20499

[6] http://www.ipswich.qld.gov.au/about_council/council_history/index.php

[7] Explore Australia, p.480, 2002, Viking

[8] Place name details (http://www.nrm.qld.gov.au/property/placenames/detail.php?id=16769)

[9] England's Oldest Town (http://www.historyfiles.co.uk/FeaturesBritain/EnglandOldestTown.htm)

[10] The Steamer "Experiment" - The Moreton Bay Courier - 20 June 1846 (http://nla.gov.au/nla.news-page540269)

[11] Ipswich City Council - Heitage Education Kit - Transport (http://www.ipswich.qld.gov.au/documents/heritage/education_kit_transport.pdf)

[12] Queensland Rail - Queensland's First Railway (http://www.history.qr.com.au/stories/)

[13] Flood Warning System for the Bremer River to Ipswich (http://www.bom.gov.au/hydro/flood/qld/brochures/brisbane_bremer/bremer.shtml)

[14] http://www.newhopecoal.com.au/about-us/history.aspx

[15] http://www.theworkshops.qm.qld.gov.au/

[16] http://www.dip.qld.gov.au/regional-planning/boonah-to-ipswich-trail.html

[17] Eagles grounded by Folau move (http://www.qt.com.au/story/2010/05/04/israel-folau-move-to-union-sucks-fans/) Chris Garry for Queensland Times 4th May 2010

[18] http://www.bremershs.eq.edu.au/

[19] http://www.rosewoodshs.qld.edu.au/

[20] http://www.sec.qld.edu.au/

[21] http://www.spcc.qld.edu.au/

[22] http://www.wcc.qld.edu.au

[23] http://woodcrestsc.eq.edu.au/wcmss/

[24] http://www.goodnajacarandafestival.com

[25] http://www.facebook.com/goodna.jacaranda/festival

[26] http://www.ipswichfestival.com.au

External links

- University of Queensland: Queensland Places: Ipswich (http://queenslandplaces.com.au/ipswich)
- City Council website (http://www.ipswich.qld.gov.au/)
- Independently run information site (http://www.ipswich-city.com/)
- Ipswich 150 Anniversary (http://www.ipswich150.com.au/proud_past/a_brief_history/index.php) - Ipswich, Queensland
- 2011 Flood Appeal (http://www.eveningstar.co.uk/news/campaigns/ipswich-united) - Ipswich United from Ipswich, Suffolk

OpenStreetMap

OpenStreetMap of Cambridge, England

URL	OpenStreetMap.org [1]
Slogan	The Free Wiki World Map
Commercial?	No
Type of site	Collaborative mapping
Registration	required for contributors
Owner	OpenStreetMap Community. Project support by OpenStreetMap Foundation.[2]
Created by	Steve Coast
Launched	1 July 2004
Alexa rank	▲ 8,085 (January 2012)[3]

OpenStreetMap (**OSM**) is a collaborative project to create a free editable map of the world. Two major driving forces behind the establishment and growth of OSM have been restrictions on use or availability of map information across much of the world and the advent of inexpensive portable GPS devices.[4]

The maps are created using data from portable GPS devices, aerial photography, other free sources or simply from local knowledge. Both rendered images and the vector dataset are available for download under a Creative Commons *Attribution-ShareAlike 2.0* licence.[5]

The OpenStreetMap approach to mapping was inspired by sites such as Wikipedia[6] ; the map display features a prominent 'Edit' tab and a full revision history is maintained. Registered users can upload GPS track logs and edit the vector data using free GIS editing tools like JOSM.[7] Various mobile applications also allow contribution of GPX tracks to the **OSM** project.

History

Steve Coast (2009)

OpenStreetMap (OSM) was founded in July 2004 by Steve Coast. In April 2006, the OpenStreetMap Foundation (OSMF) was established to encourage the growth, development and distribution of free geospatial data and provide geospatial data for anybody to use and share. In December 2006, Yahoo confirmed that OpenStreetMap could use its aerial photography as a backdrop for map production.[5]

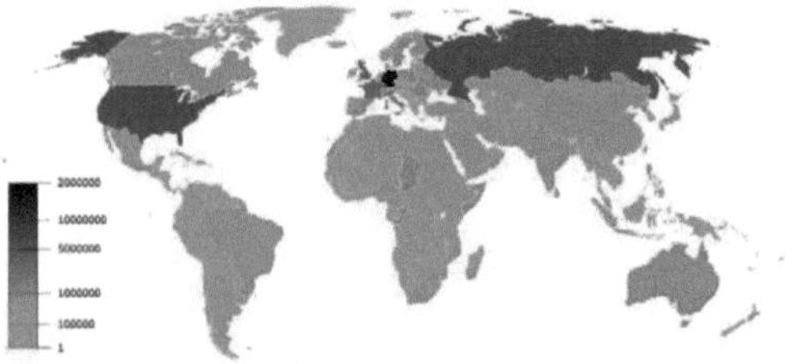

OpenStreetMap usage stats. Dark is better and popular.

In April 2007, Automotive Navigation Data (AND) donated a complete road data set for the Netherlands and trunk road data for India and China to the project[5] and by July 2007, when the first OSM international *The State of the Map* conference was held there were 9,000 registered users. Sponsors of the event included Google, Yahoo and Multimap. In August 2007 an independent project, OpenAerialMap, was launched, to hold a database of aerial photography available on open licensing[5] and in October 2007 OpenStreetMap completed the import of a US Census TIGER road dataset.[8] In December 2007 Oxford University became the first major organisation to use OpenStreetMap data on their main website.[9]

In January 2008, functionality was made available to download map data into a GPS unit for use by cyclists.[10] In February 2008 a series of workshops were held in India.[2] In March two founders announced that they have received venture capital funding of 2.4m euros for CloudMade, a commercial company that will use OpenStreetMap data.[11]

Users

By August 2008, shortly after the second *The State of the Map* conference was held, there were over 50,000 registered contributors; by March 2009 there were 100,000 and by the end of 2009 the figure was nearly 200,000. In Nov 2011, OpenStreetMap cleared 500,000 registered users.[2]

Not all registered contributors actually contribute to the map; a minority of individuals contribute the majority of the content. In March 2008 approximately 10% of the registered user base were contributing map data each month,[5] and in 2011 this percentage is around 3%.[2]

Map production

Technique

The initial map data were collected from scratch by volunteers performing systematic ground surveys using a handheld GPS unit and a notebook, digital camera, or a voice recorder. These data were then entered into the OpenStreetMap database.

More recently, the availability of aerial photography and other data sources from commercial and government sources has greatly increased the speed of this work and has allowed land-use data to be collected more accurately.

When large datasets are available, a technical team manages the conversion and import of the data.

Recording a GPS trace.

Structured ground surveys

Raw GPS data (tracklogs) for Hedge End.

Ground surveys are performed by a mapper, on foot, bicycle or in a car or boat. Map data are usually collected using a GPS unit, although this isn't strictly necessary if an area has already been traced from satellite imagery.

Once the data has been collected, they are entered into the database by uploading it on the project's website. At that point in time, no information about the kind of uploaded track is available - it could be e.g. a motorway, a footpath or a river. Thus, in a second step, editing takes place using one of several purpose-built map editors. This is usually done by the same mapper, sometimes by other users registered at OpenStreetMap.

As collecting and uploading data is separated from editing objects, contribution to the project is possible also without using a GPS unit. In particular, placing and editing objects such as schools, hospitals, taxi ranks, bus stops, pubs etc. is done based on editors' local knowledge.

Some committed contributors are systematically mapping whole towns and cities over a period of time, or organising mapping parties to intensively map a particular area over an evening or a weekend.

In addition to structured surveys, a large number of smaller edits are made by contributors to correct errors or add features.

Government data sources

Some government agencies have released official data on appropriate licenses. Much of these data have come from the United States, where the federal government does not copyright such data. See Copyright status of work by the U.S. government for more details.

- Landsat 7 satellite imagery
- Prototype Global Shorelines (PGS)
- TIGER data of the United States
- Ordnance Survey OpenData
- NRCan's CanVec vector data and GeoBase's vector data sets of Canada

Various authorities have also made more local detailed aerial photography available on suitable licenses through OpenAerialMap.

Out-of-copyright maps can be good source of information about features that don't change frequently. Copyright periods vary, but in the UK Crown copyright expires after 50 years and hence Ordnance Survey maps until the 1960s can legally be used. A complete set of UK 1 inch/mile maps from the late 1940s and early 1950s has been collected, scanned, and is available online as a resource for contributors.

Commercial data sources

Some commercial companies have donated data to the project on suitable licenses. Notably, Automotive Navigation Data (AND) provided a complete road data set for Netherlands and details of trunk roads in China and India.

In December 2006, Yahoo! confirmed that OpenStreetMap was able to make use of their vertical aerial imagery and this photography was available within the editing software as an overlay. Contributors could create their vector based maps as a derived work, released with a free and open license,[5] until the shutdown of the Yahoo! Maps API on September 13, 2011.[12]

In November 2010, Microsoft announced that the OpenStreetMap community could use Bing vertical aerial imagery as a backdrop in its editors.[5]

NearMap Pty Ltd have, since launch in November 2009, made their high-resolution PhotoMaps (currently of major Australian cities, plus some rural Australian areas) available for deriving OpenStreetMap data under a CC-BY-SA licence.[13]

Licensing

The OpenStreetMap database

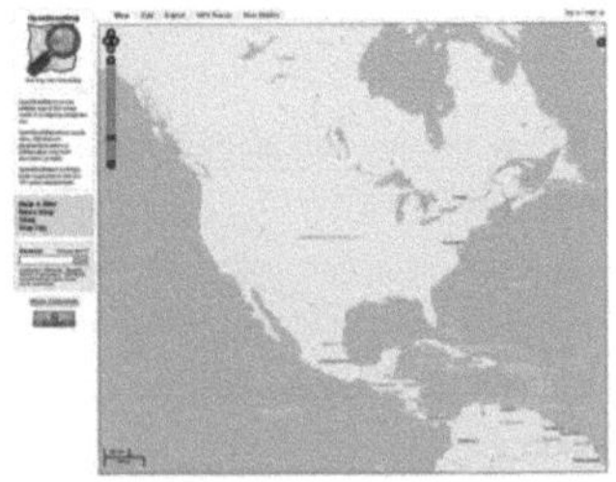

Homepage of OpenStreetMap.org

OpenStreetMap data is published under an open content license, with the intention of promoting free use and re-distribution of the data (both commercial and non-commercial). The license currently used is the Creative Commons Attribution-Share Alike 2.0 licence; however, legal investigation work and community consultation is underway to relicense the project under the Open Database License (ODbL) from Open Data Commons (ODC), claimed to be more suitable for a map data set.[5]

As part of this relicensing process, some of the map data will have to be deleted. This includes all data contributed by members who have not agreed to the new licensing terms, as well as all subsequent edits to those affected objects. It also includes any data contributed based on input data that are not compatible with the new terms. Current estimates suggest that over 95% of data will be retained in most countries, but that Australia and Poland have around a 60% acceptance rate.[14]

Input data

All data added to the project need to have a license compatible with the Creative Commons Attribution-Share Alike license. This can include out of copyright information, public domain or other licenses. All contributors must register with the project and agree to provide data on a Creative Commons BY-SA 2.0 licence, or determine that the licensing of the source data is suitable; this may involve examining licences for government data to establish whether they are compatible.

Due to the license switch, data added in future must be compatible with both the Open Database License and the new Contributor Terms in order to be accepted.

Software

Software used in the production and presentation of OpenStreetMap data is available from many different projects and each may have their own licencing. The core Ruby on Rails code to manage the databases, much user interaction and the API are available under a GNU General Public License.[2]

Certain parts of the software, such as the flash-based map editor *Potlatch2*, have been made available as public domain.[2]

Usage

OpenStreetMap data is available through an increasing number of sites and in different formats:

Online raster maps

Mapping is available online at general mapping sites:

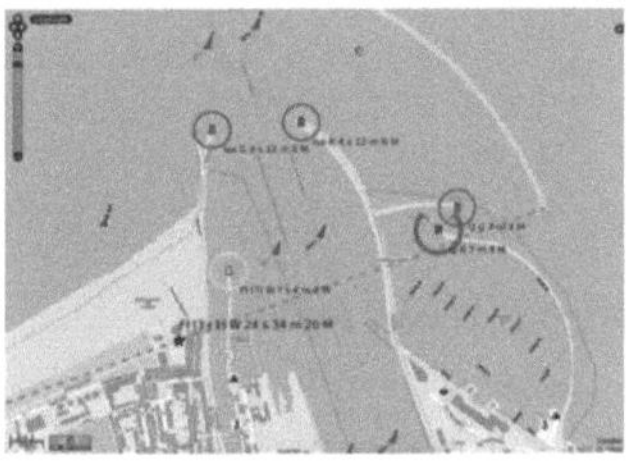

OpenSeaMap.org

Map	Theme	Area
OpenStreetMap [15]	general, cyclists, debugging	Worldwide
OpenStreetBrowser [16]	overlays, search, navigation, wikipedia, public transport	Worldwide
Information Freeway [17]	general, almost realtime	Worldwide
OSM WMS Servers [18]	general, Web Map Services	Worldwide
OpenSeaMap	nautical chart	Worldwide, multilingual: seas, oceans, waterways, weather charts, harbours, Wikipedia
FreeMap [19]	walkers	parts of the UK

Reit- und Wanderkarte [20]	walkers and riders	Austria, Germany, Switzerland
TopOSM [21]	walkers and riders	United States
OpenCycleMap [22]	cyclists	Worldwide
OpenMTBMap [23]	cyclists (esp. mountainbikes)	Worldwide
YourNavigation [24]	routing	Worldwide
OpenRouteService [25]	routing	Europe
OpenOrienteeringMap [26]	orienteering style	Worldwide
OpenPisteMap [27]	skiing	some European and USA resorts[2]
Bing OSM "Map App" [28]	general	Worldwide
CloudMade	general, mobile and various other custom styles	Worldwide
OpenAviationMap [29]	Airspace indexing and classification	Braunschweig Germany
MapQuest Open (beta) [30]	general, routing	Worldwide
NearMap	up-to-date photomaps	populated areas of Australia
OSMTransport [31]	public transport	Worldwide
ÖPNV-Karte, or OpenBusMap [32]	public transport	Worldwide
OSM Mapper [33]	Debugging maps by Ito World Ltd	
OpenLinkMap [34]	Map showing additional information, e.g. websites, wikipedia, contact, ...	
Busroutes.in [35]	public transport bus routes	India (Chennai) [Bangalore and Delhi under development stage]

Off-line raster maps

- Marble viewer by KDE for GNU/Linux, Mac OS X, and Windows
- Kosmos Renderer for Windows
- OpenSeaMap downloadable charts for PC (all operating systems), Garmin, Lowrance, PDA, PNA
- eBookMaps [36] downloadable street maps for ebook readers (Kindle, Nook etc.)

Handheld GPS receivers

- Garmin: There are several programs available that can convert OpenStreetMap data into Garmin compatible maps[2]
- Bryton Ryder 50 is the first cycling computer and SatNav device to come with OpenStreetMap data factory-loaded[37]

Mobile phones

A wide variety of programs exist by now that use OpenStreetMap data to display maps on mobile phones,[2] supporting a large number of different devices. Amongst supported devices are nearly all phones that can run Java ME, making OpenStreetMap available to the majority of current phones, as well as platforms like Maemo, Android,[2] the iPhone,[2] Windows Mobile[2] or Openmoko.[2] The various programs distinguish themselves according to key features like if they use raster maps or vector maps, need an Internet connection or can be used offline, support alternate render layers such as cycle paths,[38] support address search or advanced features like routing, thus catering for a large number of different needs for viewing maps. With the inclusion of OpenStreetMap support in several programming libraries for mobile phones[39] [40] it is also possible to include OSM maps in other location-based service.

Flickr

Yahoo has started using OpenStreetMap data within its Flickr service for various cities around the world, including Baghdad, Beijing, Kabul, Santiago, Sydney and Tokyo.[41] [42] [43]

Blog/CMS Integration

The WordPress OpenStreetMap plug-in[44] inserts maps and adds geo meta tags to posts / pages of WordPress blogs.

Drupal has multiple OpenStreetMap plug-in modules[45] to integrate database content with OpenStreetMaps.

Joomla also has a variety of OpenStreetMap related modules[46] and plug-ins.[47]

Map data

Map data is available for download in a variety of formats and for different geographical areas:

- OpenStreetMap osm[48]
- CloudMade osm, garmin[49]
- Geofabrik osm, esri[50]

OSM provides information that other map sites don't provide:

- Territorial waters
- High speed rail in China

Monopoly City Streets

In 2009, Hasbro, the games company behind the property trading board game *Monopoly*, launched *Monopoly City Streets*, a massively multiplayer online game which allowed players to "buy" streets all over the world. Although the game used map tiles from Google Maps and the Google Maps API to display the game board, the underlying vector street data was obtained from OpenStreetMap.[51] The online game was meant to be a limited time offering, closing at the end of January 2010.[52]

Streets Clock - OpenStreetMap vectors in physical product design

In 2009, Austrian mass customization company Fluid Forms [53] introduced the vectors of OpenStreetMaps into the proprietary product "Streets Clock". It is an online design interface with which people design customized wall clocks from a city's street-map. Fluid Forms uses the vectors to enable visitors of the website to get previews of how the clock will look in real-time.[54]

Haiti earthquake

During the 2010 Haiti earthquake, OpenStreetMap and Crisis Commons volunteers used available satellite imagery[55] to map the roads, buildings and refugee camps of Port-au-Prince in just two days, building "*the most complete digital map of Haiti's roads*".[56]

The resulting data and maps have been used by several organisations providing relief aid, such as the World Bank,[57] the European Commission Joint Research Centre,[58] the Office for the Coordination of Humanitarian Affairs,[59] UNOSAT[59] and others.[2]

Software

Map rendering and presentation

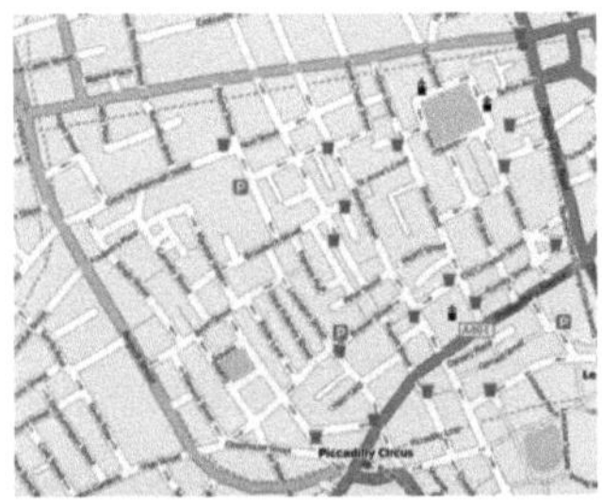

OpenStreetMap of Soho, central London (Mapnik layer)

- Online Ajax-style OpenLayers based slippy map interface. Displaying map tiles rendered by:
  - Mapnik — C++ open source map rendering software used on the main OSM site
  - Osmarender — XSLT style sheets for generating SVG images from OSM data. Used as part of the Tiles@home distributed map-rendering system for the main OSM site.
  - MapOSMatic [60] — generates city maps of using OSM data with the map split in squares allowing to easily look for streets, and an index of the streets with references to the squares on the map.
  - Cartagen [61] - javascript/HTML5 "canvas" element based renderer (client-side rendering).
- Desktop
  - CartoType [62] - Map rendering and routing library for Windows, Mac OS, and Linux (commercial).
  - osm-gps-map [63] - Open source Gtk+ widget for displaying map and track data.
  - GMap.NET [64] — Offers a Free .NET control for displaying OSM and other maps on desktop and windows mobile.
  - Maperitive — Windows and Linux application for rendering maps, it is a successor to Kosmos.
  - OSMMap [65] — Windows tool for browsing the OSM Maps (using Mapnik and Osmarender)
  - tangoGPS [66] — Open source, Linux-based mapping application for use with or without GPS. Suitable for use on many Linux platforms including desktops, the eeePC and phones such as the Openmoko Neo.
  - Prune - Open source Java application for viewing and editing data recorded by GPS. Uses OSM for the map layers.
  - GPSDrive [67] - Open source car navigation application running on Linux/BSD/OS X
  - Marble - Free and open source desktop-globe, which supports OpenStreetMap data.
  - Navit Free, modular, touch screen friendly, navigation system with GPS tracking, which can use and render OpenStreetMap data.
  - OSM2Go [68] for Maemo

- Nokia's Ovi Maps for Maemo
- Quantum GIS — Open source desktop GIS which includes OSM plugin
- luckyGPS [69] - Open source mapping,tracking and offline routing/navigation software

- Phone and GPS displays
  - AndNav2 [70] and Navit [71] for Maemo and Android
  - CartoType [62] - Map rendering and routing library for iPhone, iPad, Android and Symbian OS (commercial).
  - City Maps 2Go [72] offline satnav application for iOS (iPhone and iPad)(paid, closed source)
  - ForeverMap [73] offline satnav application for iOS (iPhone and iPad), Android (paid, closed source)
  - Galileo Offline Maps [74] offline satnav application for iOS (iPhone and iPad)(paid, closed source)
  - GpsMid [75] Java ME application
  - HotSpot Navigator [76] Java ME application
  - iLOS [77] online OSM Map and POI search for iPhone and iPad (universal)
  - MapDroyd [78] offline mapping application for Android (free, closed source)
  - Mappero [79] for Maemo
  - Maps (-) [80] offline map viewer/GPS navigator for Android
  - MapsWithMe [81] Offline map viewer for iOS (iPhone, iPad, iPod Touch) (free, closed source)
  - Mobile Street Map [82] Java ME offline map viewer (free, open source)
  - NavDroyd [83] offline satnav application for Android (paid, closed source)
  - NaviPOWM [84] offline map viewer and data collector for Windows Mobile
  - OffMaps 2 [85] offline satnav application for iOS (iPhone and iPad)(paid, closed source)
  - Opentouchmap.org [86] a web interface optimized for touch screen devices such as the iPhone
  - OruxMaps [87] - Moving-map, tracks, routes, waypoints, auto-cached maps online, offline, free Android app.
  - OsmAnd [88] online/offline mapping application for Android (free, open source)
  - OSMDroid [89] tools for Android to interact with OpenStreetMap data.
- Tablets
  - World Maps Offline [90] online/offline map viewer for iOS (iPhone, iPad, iPod Touch) (paid, closed source)
- Wordpress Blogs via OSM Plugin

Map data collection

- GPSBabel — converts tracklogs from one format to another.
- AFTrack [91] — Symbian 60v3 for Nokia mobile phones, live tracking and routing software, from version 1.03 onwards
- WhereAmI - Symbian S60, S80, and UIQ live tracking and OSM data download and rendering (GPL'ed)
- MyMotion [92] — Windows CE tracking software
- TrackMyJourney (TMJ) [93] - Java ME tracking software, which show OSM data, and uses OSM data for routing over the air.
- OSM2Go — Mobile/Desktop Hildon/Gtk editor includes tracking functionality
- OSM Live [94] - displays who is editing the Open Street Map in realtime
- iLOE [95] - (iPhone Little OpenStreetMap Editor) Editor for the iPhone (Upload and edit nodes, collecting and editing)
- MapZen POI Collector [96] — iPhone Points of Interest editor by CloudMade
- OSMTracker — track logging with quick tagging ability. It's available for Android [97] and Windows Mobile [98].

Map data editing

- Potlatch — Online Flash editor
- JOSM — Desktop Java editor
- Merkaartor — Desktop Qt editor
- OSM2Go — Mobile/Desktop Hildon/Gtk editor
- Quantum GIS — Desktop GIS which includes OSM plugin
- Vespucci - Android editor

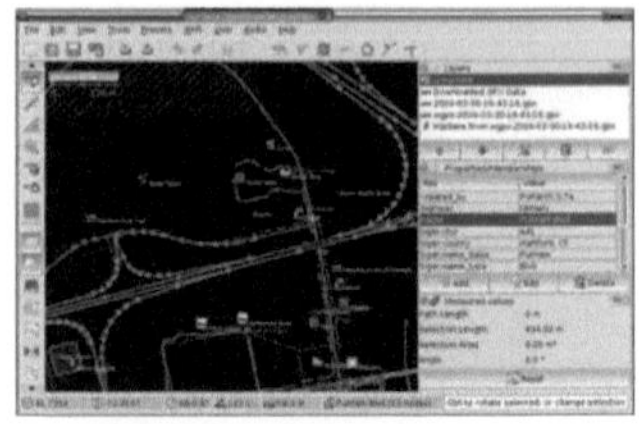

JOSM - Java-based map editor

Routing (directions, navigation)

There are lists of online and offline routers available on the OSM wiki. Following routing software supports OpenStreetMaps:

- CycleStreets (UK)
- Gosmore
- MoNav - Makes use of Contraction Hierarchies to allow for routing on continent sized graphs, even on mobile devices.
- Navit
- Pyroute
- OpenRouteService (Europe)

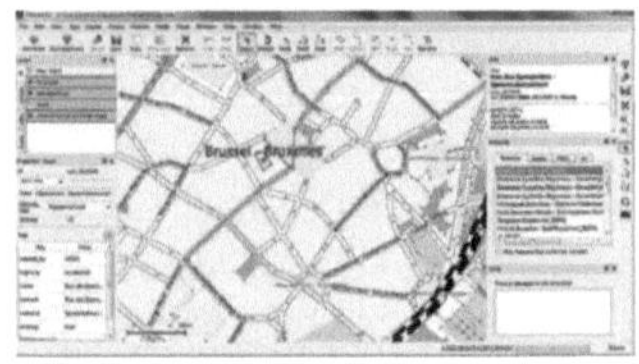

Merkaartor - Multi-platform Qt-based map editor

- Traveling salesman - Open source Java application working as route-planner and GPS-Navigator using the map-data from OpenStreetMap via LibOSM

Mobile routing

- OsmAnd (Android) Can route using CloudMade or YOURS.
- AndNav2 (Android) No longer actively developed.
- OpenSatNav (Android)
- GpsMid [75] (Java ME)
- We-Travel [99] (Java ME)

GPS tracking

- GPS-Trace Orange [100] - personal GPS tracking service, based on OSM[101]

Data format

OpenStreetMap uses a topological data structure.

- **Nodes** are points with a geographic position. Positions are stored as coordinates (pairs of a latitude and a longitude) according to WGS 84.[102]
- **Ways** are ordered lists of *nodes*, representing a polyline or polygon.
- **Relations** are groups of *nodes*, *ways* and other *relations* which can be assigned certain properties.
- **Tags** can be applied to nodes, ways or relations and consist of *name=value* pairs. They are used to store metadata about the map objects (such as their type, their name and their physical properties).

Up to version 0.4 of the protocol, an intermediate element between nodes and ways (*segments*, line segments which two node produced) was also used.

A recommended ontology of map features (the meaning of *tags*) is maintained on a wiki.

See also

- Collaborative mapping
- Counter-mapping
- Google Map Maker
- Google Maps
- Neogeography
- OpenLayers
- WikiMapia

References

[1] http://www.openstreetmap.org
[2] "FAQ" (http://wiki.openstreetmap.org/wiki/FAQ#Who_owns_OpenStreetMap.3F). *OpenStreetMap Wiki*. . Retrieved 15 April 2011.
[3] "Openstreetmap.org Site Info" (http://www.alexa.com/siteinfo/OpenStreetMap.org). Alexa Internet. . Retrieved 2012-01-02.
[4] Mark Anderson (18 Oct 2006), "Global Positioning Tech Inspires Do-It-Yourself Mapping Project" (http://news.nationalgeographic.com/news/2006/10/061018-street-maps.html), *National Geographic News*, Digital Places Special News Series, , retrieved 25 Nov 2011
[5] Coast, Steve (4 December 2006). "Yahoo! aerial imagery in OSM" (http://old.opengeodata.org/2006/12/04/yahoo-aerial-imagery-in-osm/index.html). *OpenGeoData*. . Retrieved 15 April 2011.
[6] Frederick Ramm,Jochen Topf, Steve Chilton (2011). *OpenStreetMap: Using and Enhancing the Free Map of the World*. UIT Cambridge.
[7] "Ja, ich würde OpenStreetMap mit Wikipedia vergleichen" (http://derstandard.at/1246542045463/Ja-ich-wuerde-OpenStreetMap-mit-Wikipedia-vergleichen) (in German). *Der Standard*. 13 July 2009. . Retrieved 25 October 2010.
[8] Willis, Nathan (11 October 2007). "OpenStreetMap project imports US government maps" (http://www.linux.com/archive/feature/119493). *Linux.com*. . Retrieved 16 April 2011.
[9] Batty, Peter (3 December 2007). "Oxford University using OpenStreetMap data" (http://geothought.blogspot.com/2007/12/oxford-university-using-openstreetmap.html). Geothought. . Retrieved 16 April 2011.
[10] Fairhurst, Richard (13 January 2008). "Cycle map on your GPS" (http://www.systemed.net/blog/legacy/entry080113140415.html). *Système D*. . Retrieved 16 April 2011.
[11] "We're funded!" (http://blog.cloudmade.com/2008/03/17/were-funded/). CloudMade. 17 March 2008. . Retrieved 16 April 2011.
[12] Raj Mata (13 Jun 2011), "Yahoo! Maps APIs Service Closure Announcement – New Maps Offerings Coming Soon!" (http://web.archive.org/web/20110623122827/http://developer.yahoo.com/blogs/ydn/posts/2011/06/yahoo-maps-apis-service-closure-announcement-new-maps-offerings-coming-soon/), *Yahoo! Developer Network (YDN)*, archived from the original (http://developer.yahoo.com/blogs/ydn/posts/2011/06/yahoo-maps-apis-service-closure-announcement-new-maps-offerings-coming-soon/) on 23 Jun 2011, , retrieved 25 Nov 2011
[13] "Community licence" (http://www.nearmap.com/products/community-licence). NearMap. . Retrieved 16 April 2011.
[14] Simon Poole. "OSM V1 Objects ODbL acceptance statistics" (http://odbl.poole.ch). . Retrieved 2011-12-15.
[15] http://OpenStreetMap.org
[16] http://www.openstreetbrowser.org/
[17] http://www.informationfreeway.org
[18] http://wiki.openstreetmap.org/wiki/WMS
[19] http://www.free-map.org.uk
[20] http://wanderreitkarte.de/
[21] http://toposm.com/
[22] http://www.opencyclemap.org/
[23] http://www.openmtbmap.org/
[24] http://www.yournavigation.org/
[25] http://openrouteservice.org/
[26] http://oobrien.com/oom/
[27] http://www.OpenPisteMap.org
[28] http://www.bing.com/maps/explore/#5003/s=w/5872/style=Mapnik&lat=51.512222&lon=0&z=11&pid=50735
[29] http://www-public.tu-bs.de:8080/~y0028089/OAM/
[30] http://open.mapquest.co.uk/
[31] http://www.3liz.fr/public/osmtransport/
[32] http://openbusmap.org/
[33] http://www.itoworld.com/static/osmmapper
[34] http://openlinkmap.org
[35] http://busroutes.in/

[36] http://www.ebookmaps.com
[37] Jones, Jeff (4 November 2010). "Bryton Rider 50 and Rider 30 GPS units – First look" (http://www.bikeradar.com/news/article/bryton-rider-50-and-rider-30-gps-units-first-look-28318). BikeRadar. . Retrieved 16 April 2011.
[38] "Feature-Comparison" (http://wiki.andnav.org/index.php/Feature-Comparison). *AndNavWiki*. . Retrieved 16 April 2011.
[39] "Mobile Map API SDK download" (http://www.nutiteq.com/mgmaps-lib-sdk-downloads). Nutiteq. . Retrieved 16 April 2011.
[40] "iPhone & iPad SDK" (http://cloudmade.com/products/iphone-sdk). CloudMade. . Retrieved 16 April 2011.
[41] "Around the world and back again" (http://blog.flickr.net/en/2008/08/12/around-the-world-and-back-again/). blog-flickr.net. . Retrieved 7 November 2008.
[42] "More cities" (http://blog.flickr.net/en/2008/10/30/more-cities/). blog-flickr.net. . Retrieved 7 November 2008.
[43] Tim Waters (16 Sep 2008), "Japanese progress in osm. Amazing stuff!" (http://web.archive.org/web/20110722023011/http://thinkwhere.wordpress.com/2008/09/16/japan-osm-progress/), *thinkwhere*, archived from the original (http://thinkwhere.wordpress.com/2008/09/16/japan-osm-progress/) on 22 Jul 2011, , retrieved 25 Nov 2011
[44] "WordPress OpenStreetMap plug-in" (http://wordpress.org/extend/plugins/osm/). . Retrieved 4 April 2010.
[45] "Drupal OpenStreetMap plug-in modules" (http://drupal.org/search/apachesolr_search/openstreetmap?filters=type:project_project). . Retrieved 11 August 2010.
[46] "Joomla OpenStreetMap osmMap Module" (http://extensions.joomla.org/extensions/maps-a-weather/maps-a-locations/maps/7913). . Retrieved 11 August 2010.
[47] "Joomla OpenStreetMap WISroGIS plug-in" (http://extensions.joomla.org/extensions/maps-a-weather/maps-a-locations/maps/5414). . Retrieved 11 August 2010.
[48] "planet.openstreetmap.org" (http://planet.openstreetmap.org/). OpenStreetMap. . Retrieved 15 April 2011.
[49] "CloudMade Downloads" (http://downloads.cloudmade.com/). CloudMade. . Retrieved 15 April 2011.
[50] "Download-Bereich" (http://download.geofabrik.de/osm/). Geofabrik. . Retrieved 15 April 2011.
[51] Raphael, JR (8 September 2009). "'Monopoly City Streets' Online Game: Will Buying Park Place Be Any Easier?" (http://www.pcworld.com/article/171610/monopoly_city_streets_online_game_will_buying_park_place_be_any_easier.html). *PC World*. . Retrieved 15 April 2011.
[52] (staff) (9 Sep 2009), "Monopoly game launches on Google" (http://news.bbc.co.uk/2/hi/technology/8245700.stm), *BBC News*, , retrieved 25 Nov 2011
[53] http://www.fluid-forms.com
[54] Williams, Stephen: The OpenStreetMap Streets Clock (http://fluidforms.eu/blog/2009/11/26/the-openstreetmap-streets-clock/)
[55] Forrest, Brady (1 February 2010). "Technology Saves Lives In Haiti" (http://www.forbes.com/2010/02/01/text-messages-maps-technology-breakthroughs-haiti.html). *Forbes.com*. . Retrieved 15 April 2011.
[56] "Digital Help for Haiti" (http://gadgetwise.blogs.nytimes.com/2010/01/27/digital-help-for-haiti/). *The New York Times*. 27 January 2010. . Retrieved 15 April 2011.
[57] Turner, Andrew (3 February 2010). "World Bank Haiti Situation Room - featuring OSM" (http://www.flickr.com/photos/ajturner/4329833289/in/photostream/). .
[58] "Haiti Earthquakes: Infrastructure Port-au-Prince 15/01/2010" (http://ec.europa.eu/dgs/jrc/downloads/jrc_pp_haiti_map_04.pdf) (PDF). 15 January 2010. .
[59] Batty, Peter (14 February 2010). "OpenStreetMap in Haiti - video" (http://geothought.blogspot.com/2010/02/openstreetmap-in-haiti-video.html). .
[60] http://www.maposmatic.org/
[61] http://cartagen.org/
[62] http://www.cartotype.com
[63] http://nzjrs.github.com/osm-gps-map
[64] http://greatmaps.codeplex.com
[65] http://www.mchme.com/openstreetmap_client.php
[66] http://www.tangogps.org
[67] http://www.gpsdrive.de/
[68] http://wiki.openstreetmap.org/wiki/Osm2go
[69] http://www.luckygps.com/
[70] http://www.andnav.org/
[71] http://wiki.navit-project.org/index.php/Navit_on_Android
[72] http://ulmon.com/offlinemaps
[73] http://www.skobbler.co.uk/
[74] http://galileo-app.com
[75] http://gpsmid.sourceforge.net/
[76] http://www.hsnav.net/
[77] http://eiseler.de/iphone/Johannes_iPhone_Site/iLOS.html
[78] http://www.mapdroyd.com
[79] http://www.mardy.it/mappero

[80] http://maps.coderminus.com
[81] http://www.mapswithme.com
[82] http://tengel.github.com/
[83] http://www.navdroyd.com
[84] http://sourceforge.net/projects/navipowm/
[85] http://www.offmaps.com/
[86] http://www.opentouchmap.org/
[87] http://www.oruxmaps.com/index_en.html
[88] http://www.osmand.net
[89] http://code.google.com/p/osmdroid/
[90] http://www.craftycrew.com/worldmapsoffline/
[91] http://www.afischer-online.de/sos/AFTrack/#3rd
[92] http://users.bigpond.net.au/ian_pendlebury/mymad.htm
[93] http://www.trackmyjourney.co.uk/
[94] http://datenkueche.com/osmlive
[95] http://eiseler.de/iphone/Johannes_iPhone_Site/iLOE.html
[96] http://cloudmade.com/products/mapzen-poi-collector
[97] http://code.google.com/p/osmtracker-android
[98] http://wiki.openstreetmap.org/wiki/OSMtracker
[99] http://we-travel.biz/
[100] http://gps-trace.com/
[101] "GSP-Trace Orange" (http://gps-trace.com/?page=features). GPS-trace.com. . Retrieved 30 September 2010.
[102] WGS 84 (http://wiki.openstreetmap.org/wiki/WGS_84) *OpenStreetMap Wiki*

Further reading

- Bennett, Jonathan (2010). *OpenStreetMap: Be your own Cartographer* (http://www.packtpub.com/openstreetmap/book). Packt Publishing. p. 252. ISBN 978-1847197504.
- Ramm, Frederik; Topf, Jochen; Chilton, Steve (2010). *OpenStreetMap: Using and Enhancing the Free Map of the World* (http://www.openstreetmap.info). UIT Cambridge. p. 386. ISBN 978-1-90686-011-0.
- OpenStreetMap in the media (2005 to present) (OSM wiki link (http://wiki.openstreetmap.org/wiki/OpenStreetMap_in_the_media))
- OpenStreetMap in scientific publications (OSM wiki link (http://wiki.openstreetmap.org/wiki/Research))

External links

- OpenStreetMap (http://www.openstreetmap.org/) - project page with editing interface
- OpenStreetBrowser (http://www.openstreetbrowser.org) - simple map viewer
- OpenStreetMap Foundation (http://www.osmfoundation.org/)
- OpenStreetMap Wiki (http://wiki.openstreetmap.org/wiki/Main_Page)
- *Map Compare* tools.geofabrik.de (http://tools.geofabrik.de/mc/?mt0=mapnik&mt1=googlemap) online-comparison between Google Maps and OpenStreetMap
- OpenSource navigation system (https://sourceforge.net/projects/lama/)
- Illustrated BBC report about OpenStreetMap helping in Haiti after the earthquake (http://news.bbc.co.uk/1/hi/magazine/8517057.stm)
- Evolution of European OpenStreetMap coverage 2006-2010 (video) (http://vimeo.com/16146087)
- OSM 2008: A Year of Edits (video) (http://vimeo.com/2598878)

Aerial_photography

Aerial photography is the taking of photographs of the ground from an elevated position. The term usually refers to images in which the camera is not supported by a ground-based structure. Cameras may be hand held or mounted, and photographs may be taken by a photographer, triggered remotely or triggered automatically. Platforms for aerial photography include fixed-wing aircraft, helicopters, balloons, blimps and dirigibles, rockets, kites, poles, parachutes, and vehicle mounted poles . Aerial photography should not be confused with Air-to-Air Photography, when aircraft serve both as a photo platform and subject.

The Georgian terrace of Royal Crescent (Bath, England) from a hot air balloon

London Docklands in the fog from a Eurocopter AS355

History

Aerial photography was first practiced by the French photographer and balloonist Gaspard-Félix Tournachon, known as "Nadar", in 1858 over Paris, France. [1]

The first use of a motion picture camera mounted to a heavier-than-air aircraft took place on April 24, 1909 over Rome in the 3:28 silent film short, *Wilbur Wright und seine Flugmaschine*.

The first special semiautomatic aerial camera was designed in 1911 by Russian military engineer — Colonel Potte V. F.[2] This aerial camera was used during World War I.

The use of aerial photography for military purposes was expanded during World War I by many other aviators such as Fred Zinn. One of the first notable battles was that of Neuve Chapelle.

Honoré Daumier, "Nadar élevant la Photographie à la hauteur de l'Art" (Nadar elevating Photography to Art), published in *Le Boulevard*, May 25, 1862.

Aerial mapping came into use on the battlefronts during World War I. In January 1918, General Allenby used five Australian pilots from No. 1 Squadron AFC to photograph a 624 square miles (1620 km^2) area in Palestine as an aid to correcting and improving maps of the Turkish front. Lieutenants Leonard Taplin, Allan Runciman Brown, H. L. Fraser, Edward Patrick Kenny, and L. W. Rogers photographed a block of land stretching from the Turkish front lines 32 miles (51 km) deep into their rear areas. Beginning 5 January, they flew with a fighter escort to ward off enemy fighters. Using Royal Aircraft Factory BE.12 and Martinsyde airplanes, they not only overcame enemy air attacks, but also bucked 65 mile per hour winds, antiaircraft fire, and malfunctioning equipment to complete their task circa 19 January 1918.[3]

One of the most successful pioneers of the commercial use of aerial photography was by Sherman Fairchild who started his own aircraft firm Fairchild Aircraft to develop and build specialized aircraft for high altitude aerial survey missions.[4] One Fairchild aerial survey aircraft in 1935 carried unit that combined two synchronized cameras, and each camera having five six inch lenses with a ten inch lenses and took photos from 23,000 feet. Each photo cover two hundred and twenty five square miles. One of its first government contracts was an aerial survey of New Mexico to study soil erosion.[5] A year later, Fairchild introduced a better high altitude camera with nine-lens in one unit that could take a photo of 600 square miles with each exposure from 30,000 feet. [6]

With the advent of inexpensive digital cameras, many people now take candid photographs from commercial aircraft and increasingly from general aviation aircraft on private pleasure flights.

Uses of imagery

Aerial photography is used in cartography[7] (particularly in photogrammetric surveys, which are often the basis for topographic maps), land-use planning,[7] archaeology,[7] movie production, environmental studies, surveillance, commercial advertising, conveyancing, and artistic projects. In the United States, aerial photographs are used in many Phase I Environmental Site Assessments for property analysis. Aerial photos are often processed using GIS software.

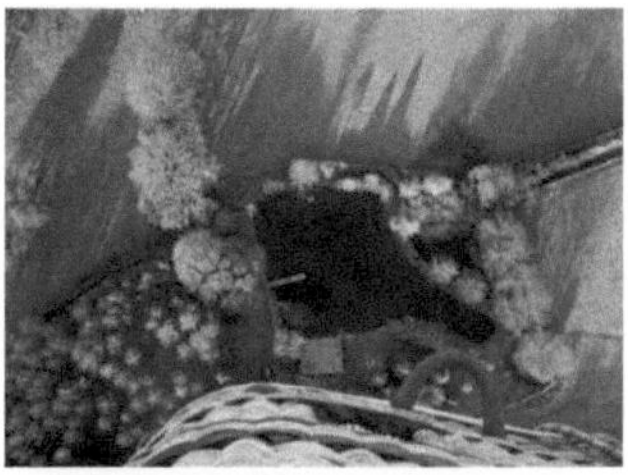

Reflection of a hot air balloon, partially obscured by a pier, an example of low-altitude aerial photography

Aerial photography platforms

Radio-controlled aircraft

Giza pyramid complex, photographed from Eduard Spelterini's balloon on November 21, 1904

Advances in radio controlled models have made it possible for model aircraft to conduct low-altitude aerial photography. This has benefited real-estate advertising, where commercial and residential properties are the photographic subject. Full-size, manned aircraft are prohibited from low flights above populated locations.[8] Small scale model aircraft offer increased photographic access to these previously restricted areas. Miniature vehicles do not replace full size aircraft, as full size aircraft are capable of longer flight times, higher altitudes, and greater equipment payloads. They are, however, useful in any situation in which a full-scale aircraft would be dangerous to operate. Examples would include the inspection of transformers atop power transmission lines and slow, low-level flight over agricultural fields, both of which can be accomplished by a large-scale radio controlled helicopter. Professional-grade, gyroscopically stabilized camera platforms are available for use under such a model; a large model helicopter with a

26cc gasoline engine can hoist a payload of approximately seven kilograms (15 lbs).

Recent (2006) FAA regulations grounding all commercial RC model flights have been upgraded to require formal FAA certification before permission to fly at any altitude in USA.

Because anything capable of being viewed from a public space is considered outside the realm of privacy in the United States, aerial photography may legally document features and occurrences on private property.[9]

Fogo island aerial shot taken from an Airbus cockpit by the pilot himself

Types of aerial photographs

Oblique photographs

Photographs taken at an angle are called *oblique photographs*. If they are taken from a low angle earth surface–aircraft, they are called *low oblique* and photographs taken from a high angle are called *high* or *steep oblique*.[10]

Vertical photographs

Vertical photographs are taken straight down.[11] They are mainly used in photogrammetry and image interpretation. Pictures that will be used in photogrammetry are traditionally taken with special large format cameras with calibrated and documented geometric properties.

Combinations

Aerial photographs are often combined. Depending on their purpose it can be done in several ways, of which a few are listed below.

- Panoramas can be made by stitching several photographs taken with one hand held camera.
- In pictometry five rigidly mounted cameras provide one vertical and four low oblique pictures that can be used together.
- In some digital cameras for aerial photogrammetry images from several imaging elements, sometimes with separate lenses, are geometrically corrected and combined to one image in the camera.

Orthophotos

Vertical photographs are often used to create orthophotos, photographs which have been geometrically "corrected" so as to be usable as a map. In other words, an orthophoto is a simulation of a photograph taken from an infinite distance, looking straight down to nadir. Perspective must obviously be removed, but variations in terrain should also be corrected for. Multiple geometric transformations are applied to the image, depending on the perspective and terrain corrections required on a particular part of the image.

Orthophotos are commonly used in geographic information systems, such as are used by mapping agencies (e.g. Ordnance Survey) to create maps. Once the images have been aligned, or "registered", with known real-world coordinates, they can be widely deployed.

Large sets of orthophotos, typically derived from multiple sources and divided into "tiles" (each typically 256 x 256 pixels in size), are widely used in online map systems such as Google Maps. OpenStreetMap offers the use of similar orthophotos for deriving new map data. Google Earth overlays orthophotos or satellite imagery onto a digital elevation model to simulate 3D landscapes.

Aerial video

With advancements in video technology, aerial video is becoming more popular. Orthogonal video is shot from aircraft mapping pipelines, crop fields, and other points of interest. Using GPS, video may be embedded with meta data and later synced with a video mapping program.

This "Spatial Multimedia" is the timely union of digital media including still photography, motion video, stereo, panoramic imagery sets, immersive media constructs, audio, and other data with location and date-time information from the GPS and other location designs.

Aerial videos are emerging Spatial Multimedia which can be used for scene understanding and object tracking. The input video is captured by low flying aerial platforms and typically consists of strong parallax from non-ground-plane structures. The integration of digital video, global positioning systems (GPS) and automated image processing will improve the accuracy and cost-effectiveness of data collection and reduction. Several different aerial platforms are under investigation for the data collection.

See also

- Aerial archaeology
- Aerial landscape art
- Aerial photographers category
- Airborne Real-time Cueing Hyperspectral Enhanced Reconnaissance
- Astrocam
- Aviation photography
- Battle of Neuve Chapelle
- Douglas Douglas-Hamilton, 14th Duke of Hamilton 1932 photo flight over Mount Everest
- Fairchild K-20 An early aerial camera
- Federal Aviation Regulations
- Geoinformatics
- Kite aerial photography
- National Monuments Record the public archive of English Heritage, who hold one of the largest collections of aerial photographs of England
- Oracle model photographic rocket
- Pictometry
- Pigeon photography
- Photogrammetry
- Remote sensing
- Satellite imagery
- TopoFlight
- Unmanned aerial vehicle

References

[1] History of Aerial Photography (http://www.papainternational.org/history.html) Professional Aerial Photographers Association (retrieved December 21, 2007)

[2] Great Soviet Encyclopedia, Article *"Аэрофотоаппарат"* (eng. *aerial camera*)

[3] http://www.southsearepublic.org/2002_1999/afc_aces_taplin.htm#top Retrieved 24 February 2011.

[4] *The Complete Encyclopedia of World Aircraft* pg. 382 ISBN 0-7607-0592-5 printed 1997

[5] "Wide Area Is Mapped From Air By Giant Ten Lens Camera" *Popular Mechanics*, October 1935 (http://books.google.com/books?id=wN8DAAAAMBAJ&pg=PA535&dq=Popular+Science+1935+plane+"Popular+Mechanics"&hl=en&ei=aJgvTv3XKdSgsQLVyIRv&sa=X&oi=book_result&ct=result&resnum=9&sqi=2&ved=0CEkQ6AEwCA#v=onepage&q&f=true) editors have stated Fairchild Aircraft in hand written comment to left of archived article

[6] "Nine Lens Aerial Camera Films 600 Square Miles" *Popular Mechanics*, April 1936 (http://books.google.com/books?id=lNsDAAAAMBAJ&pg=PA571&dq=Popular+Science+1936+plane+"Popular+Mechanics"&hl=en&ei=YQxKTqCgIeSDsgK6xpzSCA&sa=X&oi=book_result&ct=result&resnum=1&sqi=2&ved=0CCoQ6AEwAA#v=onepage&q=Popular Science 1936 plane "Popular Mechanics"&f=true)

[7] "Aerial Photography and Remote Sensing" (http://www.colorado.edu/geography/gcraft/notes/remote/remote_f.html). *University of Colorado Boulder*. 2011 [last update]. . Retrieved March 25, 2011.

[8] United States Federal Aviation Regulations FAR part 91 section 119(14CFR91.119)

[9] *California v. Ciraolo*, 476 U.S. 206 (1986)

[10] "Lecture 6.1: Classification of Photographs" (http://userpages.umbc.edu/~tbenja1/umbc7/santabar/vol1/lec6/6-1.html). *The Remote Sensing Core Curriculum*. University of Maryland, Baltimore County. 1999 [last update]. . Retrieved 2011-03-25.

[11] Short, Nicholas (2010-04-28). "Elements of Aerial Photography" (http://rst.gsfc.nasa.gov/Sect10/Sect10_1.html). *Remote Sensing Tutorial Page 10-1*. NASA. . Retrieved 2011-03-25.

External links

- Aerial photography and remote sensing (http://www.colorado.edu/geography/gcraft/notes/remote/remote_f.html)
- Air photo interpretation (http://airphotos.nrcan.gc.ca/photos101/photos101_info_e.php)
- California Coastal Records Project (http://www.californiacoastline.org/)
- Bird's eye view of the Delaware Valley (http://www.hagley.org/library/exhibits/dallinexhibit/) by the Dallin Aerial Survey Company (1924–1941)
- Historic aerial photos of Columbia, South Carolina (http://www.sc.edu/library/digital/collections/aerials/scaerialscolahome.html) from the University of South Carolina Library
- Historical Aerial Photographs of New York (http://aerial-ny.library.cornell.edu), Cornell University Library & Cornell Institute for Resource Information Sciences
- National Collection of Aerial Photography (http://aerial.rcahms.gov.uk/) The official archive of British Government declassified aerial photography.

Digital_elevation_model

A **digital elevation model** is a digital model or 3-D representation of a terrain's surface — commonly for a planet (including Earth), moon, or asteroid — created from terrain elevation data.

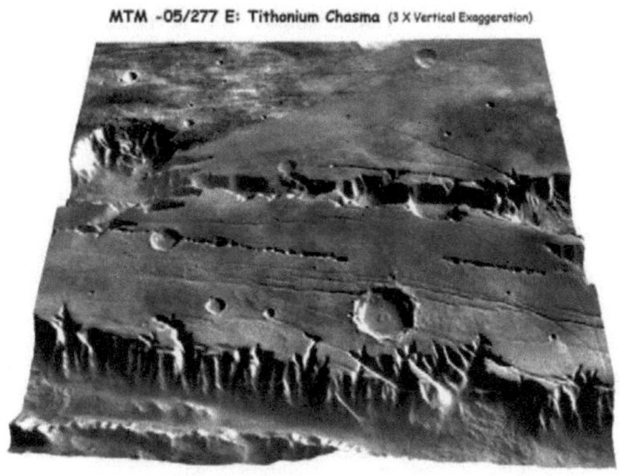

3D rendering of a DEM of Tithonium Chasma on Mars

There is no common usage of the terms **digital elevation model (DEM)**, **digital terrain model (DTM)** and **digital surface model (DSM)** in scientific literature. In the most cases the term **digital surface model** represents the earth's surface and includes all objects on it. In contrast to a DSM, the **digital terrain model** represents the bare ground surface without any objects like plants and buildings (see Figure on the right).[1] [2]

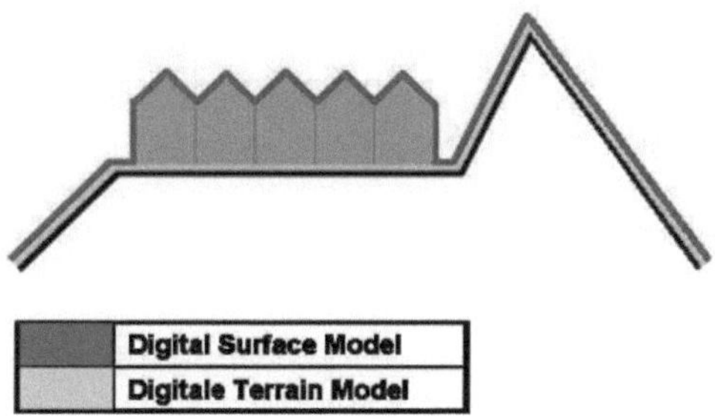

Surfaces represented by a Digital Surface Model include buildings and other objects. Digital Terrain Models represent the bare ground.

The term Digital Elevation Model is often used as a generic term for DSMs and DTMs, only representing height information without any further definition about the surface[3] . Other definitions equalise the terms DEM and DTM[4] , or define the DEM as a subset of the DTM, which is also representing other morphological elements[5] . There are also definitions which equalise the terms DEM and DSM[6] . In the Web definitions can be found which define the DEM as a digital regularly spaced GRID and a DTM as a real three-dimensional model (TIN)[7] . Most of the data providers (USGS, ERSDAC, CGIAR) use the term DEM as a generic term for DSMs and DTMs. All datasets which are captured with satellites, airplanes or other flying platforms are originally DSMs (like SRTM or the ASTER GDEM). It is possible to compute a DTM from high resolution DSM datasets with complex algorithms (Li et al, 2005). In the following the term DEM is used as a generic term for DSMs and DTMs.

A DEM can be represented as a raster (a grid of squares, also known as a heightmap when representing elevation) or as a vector-based triangular irregular network (TIN). The TIN DEM dataset is also referred as a primary (measured) DEM, whereas the Raster DEM is referred as a secondary (computed) DEM[8] . DEMs are commonly built using remote sensing techniques, but they may also be built from land surveying. DEMs are used often in geographic information systems, and are the most common basis for digitally-produced relief maps. The DEM could be acquired through techniques such as photogrammetry, LiDAR, IfSAR, land surveying, etc. (Li et al. 2005). While a DSM may be useful for landscape modeling, city modeling and visualization applications, a DTM is often required for flood or drainage modeling, land-use studies, geological applications, and much more.[9]

Production

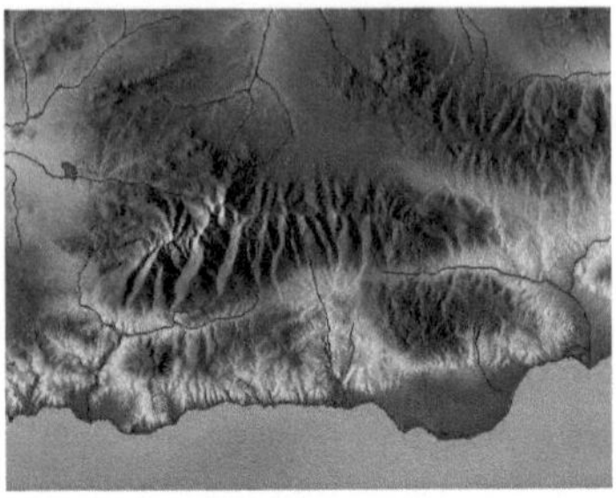

Relief map Sierra Nevada

Mappers may prepare digital elevation models in a number of ways, but they frequently use remote sensing rather than direct survey data. One powerful technique for generating digital elevation models is interferometric synthetic aperture radar: two passes of a radar satellite (such as RADARSAT-1 or TerraSAR-X), or a single pass if the satellite is equipped with two antennas (like the SRTM instrumentation), suffice to generate a digital elevation map tens of kilometers on a side with a resolution of around ten meters . Alternatively, other kinds of stereoscopic pairs can be employed using the digital image correlation method, where two optical images acquired with different angles taken from the same pass of an airplane or an Earth Observation Satellite (such as the HRS instrument of SPOT5 or the VNIR band of ASTER).[10]

In 1986, the SPOT 1 satellite provided the first usable elevation data for a sizeable portion of the planet's landmass, using two-passes stereoscopic correlation. Later, further data were provided by the European Remote-Sensing Satellite (ERS) using the same method, the Shuttle Radar Topography Mission using single-pass SAR and the ASTER instrumentation on the Terra satellite using double-pass stereo pairs.[10]

Older methods of generating DEMs often involve interpolating digital contour maps that may have been produced by direct survey of the land surface; this method is still used in mountain areas, where interferometry is not always satisfactory. Note that the contour line data or any other sampled elevation datasets (by GPS or ground survey) are not DEMs, but may be considered digital terrain models. A DEM implies that elevation is available continuously at each location in the study area.

The quality of a DEM is a measure of how accurate elevation is at each pixel (absolute accuracy) and how accurately is the morphology presented (relative accuracy). Several factors play an important role for quality of DEM-derived products:

- terrain roughness;
- sampling density (elevation data collection method);
- grid resolution or pixel size;
- interpolation algorithm;
- vertical resolution;
- terrain analysis algorithm;

Methods for obtaining elevation data used to create DEMs

- LIDAR
- Stereo photogrammetry from aerial surveys
- Real Time Kinematic GPS
- Topographic maps
- Theodolite or total station
- Doppler radar
- Focus variation
- Inertial surveys

Uses

Common uses of DEMs include:

- Extracting terrain parameters
- Modeling water flow or mass movement (for example avalanches and landslides)
- Creation of relief maps
- Rendering of 3D visualizations.
- 3d flight planning
- Creation of physical models (including raised-relief maps)
- Rectification of aerial photography or satellite imagery.
- Reduction (terrain correction) of gravity measurements (gravimetry, physical geodesy).
- Terrain analyses in geomorphology and physical geography
- Geographic Information Systems (GIS)
- Engineering and infrastructure design
- Global positioning systems (GPS)
- Line-of-sight analysis
- Base mapping
- Flight simulation
- Precision farming and forestry
- Surface analysis
- Intelligent transportation systems (ITS)
- Auto safety / Advanced Driver Assistance Systems (ADAS)

Bezmiechowa airfield 3D Digital Surface Model obtained using Pteryx UAV flying 200m above hilltop

Digital Surface Model of motorway interchange construction site. Note that tunnels are closed.

Sources

A free DEM of the whole world called GTOPO30 (30 arcsecond resolution, approx. 1 km) is available, but its quality is variable and in some areas it is very poor. A much higher quality DEM from the Advanced Spaceborne Thermal Emission and Reflection Radiometer (ASTER) instrument of the Terra satellite is also freely available for 99% of the globe, and represents elevation at a 30 meter resolution. A similarly high resolution was previously only available for the United States territory under the Shuttle Radar Topography Mission (SRTM) data, while most of the rest of the planet was only covered in a 3 arc-second resolution (around 90 meters). The limitation with the GTOPO30 and SRTM datasets is that they cover continental landmasses only, and SRTM does not cover the polar regions and has mountain and desert no data (void) areas. SRTM data, being derived from radar, represents the elevation of the first-reflected surface — quite often tree tops. So, the data are not necessarily representative of the ground surface, but the top of whatever is first encountered by the radar. Submarine elevation (known as bathymetry) data is generated using ship-mounted depth soundings. The SRTM30Plus dataset (used in NASA World Wind) attempts to combine GTOPO30, SRTM and bathymetric data to produce a truly global elevation model.[11] A novel global DEM of postings lower than 12m and a height accuracy of less than 2m is expected being generated by the TanDEM-X satellite mission which started in July 2010.

The most usual grid (raster) is between 50 and 500 meters. In gravimetry e.g., the primary grid may be 50 m, but is switched to 100 or 500 meters in distances of about 5 or 10 kilometers.

Many national mapping agencies produce their own DEMs, often of a higher resolution and quality, but frequently these have to be purchased, and the cost is usually prohibitive to all except public authorities and large corporations. DEMs are often a product of National LIDAR Dataset programs.

Free DEMs are also available for Mars: the MEGDR, or Mission Experiment Gridded Data Record, from the Mars Global Surveyor's Mars Orbiter Laser Altimeter (MOLA) instrument; and NASA's Mars Digital Terrain Model (DTM).[12]

United States

The US Geological Survey produces the National Elevation Dataset, a seamless DEM for the contiguous United States, Hawaii and Puerto Rico based on 7.5' topographic mapping. As of the beginning of 2006, this replaces the earlier DEM tiled format (one DEM per USGS topographic map).[13] [14]

See also

- TerraSAR-X:a German Earth observation satellite
- TanDEM-X: generation of a world-wide, consistent, timely, high-precision Digital Elevation Model

References

[1] "Intermap Digital Surface Model: accurate, seamless, wide-area surface models" (http://www.intermap.com/en-us/nextmap/digitalsurfacemodel.aspx). .

[2] Li, Z., Zhu, Q. and Gold, C. (2005): title=Digital terrain modeling: principles and methodology|. CRC Press. Boca Raton.

[3] Peckham, Robert Joseph; Jordan, Gyozo (Eds.)(2007): Development and Applications in a Policy Support Environment Series: Lecture Notes in Geoinformation and Cartography. Heidelberg.

[4] Podobnikar, Tomaz (2008). "Methods for visual quality assessment of a digital terrain model" (http://sapiens.revues.org/index738.html). *S.a.p.i.en.s.* **1** (2). .

[5] "DIN Standard 18709-1" (http://www.din.de/cmd?level=tpl-home&languageid=en). .

[6] Adrian W. Graham,Nicholas C. Kirkman,Peter M. Paul (2007): Mobile radio network design in the VHF and UHF bands: a practical approach. West Sussex.

[7] "Landslide Glossary USGS" (http://landslides.usgs.gov/learning/glossary.php#d). .

[8] RONALD TOPPE (1987): Terrain models — A tool for natural hazard Mapping. In: Avalanche Formation, Movement and Effects (Proceedings of the Davos Symposium, September 1986). IAHS Publ. no. 162,1987

[9] "Appendix A – Glossary and Acronyms" (http://www.environment-agency.gov.uk/commondata/acrobat/app_a_1243533.pdf). *Severn Tidal Tributaries Catchment Flood Management Plan – Scoping Stage*. UK: Environment Agency. .

[10] Nikolakopoulos, K. G.; Kamaratakis, E. K; Chrysoulakis, N. (10 November 2006). "SRTM vs ASTER elevation products. Comparison for two regions in Crete, Greece" (http://www.iacm.forth.gr/_docs/pubs/4/Nikolakopoulos_et_al_2006.pdf). *International Journal of Remote Sensing* **27** (21). ISSN 4819–4838. . Retrieved June 22, 2010.

[11] see Martin Gamache's paper on free sources of global data, http://www.terrainmap.com/downloads/Gamache_final_web.pdf

[12] A basic guide for using Digital Elevation Models with Terragen (http://www2.cs.uh.edu/~somalley/DemTutorial/#DEM)

[13] http://eros.usgs.gov/DEMNotice.html

[14] see Herbert Glarner's paper on using USGS data, http://herbert.gandraxa.com/herbert/dem.asp

- Paula Messina. "Terrain Analysis Home Page" (http://www.geo.hunter.cuny.edu/terrain). Spatial Analysis and Remote Sensing Lab at Hunter College. Retrieved 2007-02-16.
- Wilson, J.P.; Gallant, J.C. (2000). "Chapter 1" (http://media.wiley.com/product_data/excerpt/85/04713218/0471321885.pdf). In Wilson, J.P., and Gallant, J.C. (Eds.). *Terrain Analysis: Principles and Applications*. New York: Wiley. pp. 1–27. ISBN 0471321885. Retrieved 2007-02-16.

DEM file formats

- USGS DEM
- SDTS DEM
- DTED
- DIMAP (http://www.spotimage.fr/web/en/235-the-dimap-format.php,)

External links

- Maps-For-Free.com (http://www.maps-for-free.com) Free global relief maps
- Geo-Spatial Data Acquisition Homepage (http://www.emrl.byu.edu/gsda)
- GTOPO30 Homepage (http://edcdaac.usgs.gov/gtopo30/gtopo30.html)
- GTOPO30 FTP Server (ftp://edcftp.cr.usgs.gov/pub/data/gtopo30/global/)
- SRTM Homepage (http://srtm.usgs.gov/)
- SRTM30 Plus Homepage (http://topex.ucsd.edu/WWW_html/srtm30_plus.html)
- Terrainmap Homepage (http://www.terrainmap.com/)
- More information about available DEM data (http://www.vterrain.org/Elevation/index.html)
- More information about DEM by Spot Image (http://www.spotimage.fr/web/en/807-spot-3d.php)
- DEMs with TanDEM-X (http://www.terrasar.de/tandem-x_dem)
- TanDEM-X Science Home (http://www.dlr.de/HR/tdmx)

Brisbane

Brisbane

Queensland

The Brisbane skyline from Kangaroo Point

Brisbane

Population:	2,043,185 (2010)[1] (3rd)
• Density:	918/km² (2377.6/sq mi) (2006)[2]

Established:	1824
Coordinates:	27°28′22″S 153°01′40″E
Area:	5904.8 km² (2279.9 sq mi) [3]
Time zone:	AEST (UTC+10)
Location:	• 931 km (578 mi) N of Sydney • 1692 km (1051 mi) NNE of Melbourne • 1969 km (1223 mi) NE of Adelaide • 4262 km (2648 mi) E of Perth • 1450 km (901 mi) NNE of Canberra
LGA:	• Brisbane • Ipswich • Logan • Moreton Bay • Redland • Scenic Rim (part)
Region:	South East Queensland
County:	Stanley
State District:	various (38)
Federal Division:	• Blair, Bonner, Bowman, Brisbane • Dickson, Fadden, Forde, Griffith • Lilley, Longman, Moreton, Oxley • Petrie, Ryan

Mean max temp	Mean min temp	Annual rainfall
26.4 °C 80 °F	16.2 °C 61 °F	965.7 mm 38 in

Brisbane (/ˈbrɪzbən/)[4] is the capital and most populous city in the Australian state of Queensland and the third most populous city in Australia. Brisbane's metropolitan area has a population of over 2 million, and the South East Queensland urban conurbation, centred around Brisbane, encompasses a population of more than 3 million.[5] The Brisbane central business district stands on the original European settlement and is situated inside a bend of the Brisbane River approximately 23 kilometres from its mouth at Moreton Bay. The metropolitan area extends in all directions along the floodplain of the Brisbane River valley between the bay and the Great Dividing Range. While the metropolitan area is governed by several municipalities, a large proportion of central Brisbane is governed by the Brisbane City Council which is Australia's largest Local Government Area by population. The demonym of Brisbane is Brisbanian, but is known to be Brisbanite.[6]

Brisbane is named after the river on which it sits which, in turn, was named after Scotsman Sir Thomas Brisbane, the Governor of New South Wales from 1821 to 1825. The first European settlement in Queensland was a penal colony at Redcliffe, 28 kilometres (17 mi) north of the Brisbane central business district, in 1824. That settlement was soon abandoned and moved to North Quay in 1825. Free settlers were permitted from 1842. Brisbane was chosen as the capital when Queensland was proclaimed a separate colony from New South Wales in 1859.

The city played a central role in the Allied campaign during World War II as the South West Pacific headquarters for General Douglas MacArthur. Brisbane has hosted many large cultural and sporting events including the 1982 Commonwealth Games, World Expo '88 and the final Goodwill Games in 2001. Brisbane is the largest economy between Sydney and Singapore and in 2008 it was classified as a Beta– world city in the World Cities Study Group's inventory by Loughborough University.[7] It was also rated the 16th most livable city in the world in 2009 by The

Economist.

History

Prior to European settlement, the Brisbane area was inhabited by the Turrbal and Jagera people.[8] They knew the area that is now the central business district as Mian-jin, meaning "place shaped as a spike".[9] The Moreton Bay area was initially explored by Matthew Flinders. On 17 July 1799, Flinders landed at what is now known as Woody Point, which he named "Red Cliff Point", after the red-coloured cliffs visible from the bay.[10] In 1823, Governor of New South Wales, Thomas Brisbane, instructed that a new northern penal settlement be developed, and an exploration party led by John Oxley further explored Moreton Bay.[11]

Oxley discovered, named and explored the Brisbane River as far as Goodna, 20 kilometres (12 mi) upstream from the Brisbane central business district.[11] Oxley recommended Red Cliff Point for the new colony, reporting that ships could land at any tide and easily get close to the shore.[12] The party settled in Redcliffe on 13 September 1824, under the command of Lieutenant Henry Miller with 14 soldiers; some with wives and children; and 29 convicts. However, this settlement was abandoned after a year, and the colony was moved to a site on the Brisbane River now known as North Quay, 28 kilometres (17 mi) south, that offered a more reliable water supply. Chief Justice Forbes gave the new settlement the name of Edenglassie before it was named Brisbane.[13] Non-convict European settlement of the Brisbane region commenced in 1838.[14] German missionaries settled at Zions Hill, Nundah, as early as 1837, five years before Brisbane was officially declared a free settlement. The band consisted of two ministers, Christopher Eipper (1813–1894) and Carl Wilhelm Schmidt, and lay missionaries Haussmann, Johann Gottried Wagner, Niquet, Hartenstein, Zillman, Franz, Rode, Doege and Schneider.[15] They were allocated 260 hectares and set about establishing the mission, which became known as German Station.[16]

RAAF recruits marching along Queen Street, August 1940.

Free settlers entered the area over the following five years and by the end of 1840 Robert Dixon began work on the first plan of Brisbane Town in anticipation of future development.[17] Queensland was proclaimed a separate colony on 6 June 1859,[18] with Brisbane chosen as its capital, although it was not incorporated as a city until 1902. Over twenty small municipalities and shires were amalgamated in 1925, to form the City of Brisbane which is governed by the Brisbane City Council.[19] [20] 1930 was a significant year for Brisbane, with the completion of Brisbane City Hall, then the city's tallest building and the Shrine of Remembrance, in ANZAC Square, which has become Brisbane's main war memorial.[21] These historic buildings along with the Story Bridge, opened in 1940[22] are key landmarks that help define the character of the city.

During World War II, Brisbane became central to the Allied campaign when the AMP Building (now called MacArthur Central) was used as the South West Pacific headquarters for General Douglas MacArthur, chief of the Allied Pacific forces, until his headquarters were moved to Hollandia in August 1944. MacArthur had previously rejected use of the University of Queensland complex as his headquarters, as the distinctive bends in the river at St Lucia could have aided enemy bombers. Also used as a headquarters by the American troops during World War II was the T & G Building.[23] Approximately 1 million US troops passed through Australia during the war, as the primary coordination point for the South West Pacific.[24] In 1942 Brisbane was the site of a violent clash between visiting US military personnel and Australian servicemen and civilians which resulted in one death and several injuries. This incident became known colloquially as the Battle of Brisbane.[25]

Postwar Brisbane had developed a "big country town" stigma, an image the city's politicians and marketers were very keen to remove.[26] In the late 1950s an anonymous poet known as The Brisbane Bard generated much attention on the city which helped to shake this stigma.[27] [28] Despite steady growth, Brisbane's development was punctuated by infrastructure problems. The State government under Joh Bjelke-Petersen began a major program of change and urban renewal, beginning with the central business district and inner suburbs. Trams in Brisbane were a popular mode of public transport, until the network was closed in 1969, leaving Melbourne as the last Australian city to operate a tram network. The 1974 Brisbane flood was a major disaster which temporarily crippled the city. During this era, Brisbane grew and modernised rapidly becoming a destination of interstate migration. Some of Brisbane's popular landmarks were lost, including the Bellevue Hotel in 1979 and Cloudland in 1982, demolished in controversial circumstances by the Deen Brothers demolition crew. Major public works included the Riverside Expressway, the Gateway Bridge, and later, the redevelopment of South Bank, starting with the Queensland Art Gallery.

Brisbane hosted the 1982 Commonwealth Games and the 1988 World Exposition (known locally as World Expo 88). These events were accompanied by a scale of public expenditure, construction and development not previously seen in the state of Queensland.[29] [30] Brisbane's population growth has exceeded the national average every year since 1990 at an average rate of around 2.2% per year. Brisbane was again hit by a major flood in January 2011. The Brisbane River did not reach the same height as the previous 1974 flood but still caused extensive damage and disruption to the city.[31] [32]

The Old Windmill in Wickham Park in Brisbane. Built by convicts in 1824, it is one of Brisbane's most significant and oldest heritage landmarks.

ANZAC Square and the Shrine of Remembrance, Brisbane (c1930).

American Fleet Marching Down Queen Street, March 1941.

World Expo 88 as seen from the Brisbane River.

Queensland's first Government House was completed in 1862 and is located within the grounds of the Queensland University of Technology near the City Botanic Gardens

Geography

Brisbane's central business district, viewed from Kangaroo Point, lies on a curve of the Brisbane river.

Brisbane is in the southeast corner of Queensland, Australia. The city is centred along the Brisbane River, and its eastern suburbs line the shores of Moreton Bay. The greater Brisbane region is on the coastal plain east of the Great Dividing Range. Brisbane's metropolitan area sprawls along the Moreton Bay floodplain from Caboolture in the north to Beenleigh in the south, and across to Ipswich in the south west.

The city of Brisbane is based on a hilly surface.[33] The urban area, including the central business district, are partially elevated by spurs of the Herbert Taylor Range, such as the summit of Mount Coot-tha, reaching up to 300 metres (980 ft) and the smaller Enoggera Hill. Other prominent rises in Brisbane are Mount Gravatt and nearby Toohey Mountain. Mount Petrie at 170 metres (560 ft) and the lower rises of Highgate Hill, Mount Ommaney, Stephens Mountain and Whites Hill are dotted across the city.

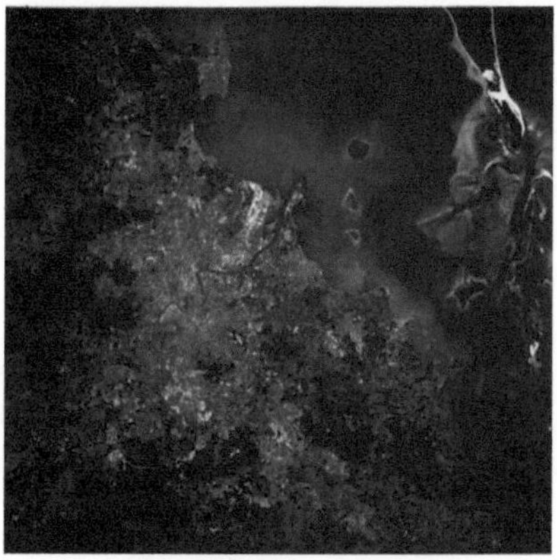

Satellite image of Brisbane Metropolitan Area.

The city is on a low-lying floodplain.[34] Many suburban creeks criss-cross the city, increasing the risk of flooding. The city has suffered three major floods since colonisation, in February 1893, January 1974, and January 2011. The 1974 Brisbane flood occurred partly as a result of "Cyclone Wanda". Heavy rain had fallen continuously for three weeks before the Australia Day weekend flood (26 – 27 January 1974).[35] The flood damaged many parts of the city, especially the suburbs of Oxley, Bulimba, Rocklea, Coorparoo, Toowong and New Farm. The City Botanic gardens were inundated, leading to a new colony of mangroves forming in the City Reach of the Brisbane River.[36]

Urban structure

The Brisbane cityscape at night, with surrounding suburbs in the foreground.

The Brisbane central business district (CBD) lies in a curve of the Brisbane river. The CBD covers only 2.2 km^2 (0.8 sq mi) and is walkable. Central streets are named after members of the royal family. Queen Street is Brisbane's traditional main street. Streets named after female members (Adelaide, Alice, Ann, Charlotte, Elizabeth, Margaret, Mary) run parallel to Queen Street and Queen Street Mall (named in honour of Queen Victoria) and perpendicular to streets named after male members (Albert, Edward, George, William). The city has retained some heritage buildings dating back to 1820's. The Old Windmill, in Wickham Park, built by convict labour in 1824,[37] [38] is the oldest surviving building in Brisbane. The Old Windmill was originally used for the grinding of grain and a punishment for the convicts who manually operated the grinding mill. The Old Windmill tower's other significant claim to fame, largely ignored, is that the first television signals in the southern hemisphere were transmitted from it by experimenters in April 1934 – long before TV commenced in most places. These experimental TV broadcasts continued until World War II.[37]

Medium density housing along the Brisbane River at Newstead.

The Old Commissariat Store, on William Street, built by convict labour in 1828, was originally used partly as a grainhouse, has also been a hostel for immigrants and used for the storage of records. Built with Brisbane tuff from the nearby Kangaroo Point Cliffs and sandstone from a quarry near today's Albion Park Racecourse, it is now the home of the Royal Historical Society of Brisbane. It contains a museum and can also be hired for small functions.[39] [40] [41] The city has a density of 379.4 people per square kilometre, which is high for an Australian city and comparable to that of Sydney. However like many western cities, Brisbane sprawls into the greater metropolitan area. This results from the fact that most of Brisbane's housing stock consists of detached houses.

Early legislation decreed a minimum size for residential blocks resulting in few terrace houses being constructed in Brisbane. Recently the density of the city and inner city neighbourhoods has increased with the construction of apartments, with the result that the population of the central business district has doubled over the last 5 years.[42] The high density housing that historically existed came in the form of miniature Queenslander-style houses which resemble the much larger traditional styles but are sometimes only one quarter the size. These miniature Queenslanders are becoming scarce but can still be seen in the inner city suburbs. Multi residence accommodations (such as apartment blocks) are relatively new to Brisbane, with few such blocks built before 1970, other than in inner suburbs such as New Farm. Pre-1950 housing was often built in a distinctive architectural style known as a Queenslander, featuring timber construction with large verandahs and high ceilings. The relatively low cost of timber in South-East Queensland meant that until recently most residences were constructed of timber, rather than brick or stone. Many of these houses are elevated on stumps (also called "stilts"), that were originally timber, but are now frequently replaced by steel or concrete. Brisbane is home to several of Australia's tallest buildings. Brisbane's tallest buildings are Soleil at 243 metres, Aurora Tower at 207 metres, Riparian Plaza at 200 metres, One One One Eagle Street at 195 metres, and Infinity, at 249 metres, which is currently under construction.[43]

1	Walter Taylor Bridge (road) (left), Albert Bridge (rail) (centre), unnamed bridge (rail) (right), Jack Pesch Bridge (far right)
2	Eleanor Schonell Bridge (Green Bridge) (pedestrians, pedal cycles, buses)
3	Merivale Bridge (rail)
4	William Jolly Bridge (road)
5	Victoria Bridge
6	Captain Cook Bridge
7	Story Bridge
8	Pacific Motorway
9	Suncorp Stadium (Lang Park) (Rugby league/Rugby Union/Soccer ground)
10	Norman Creek (Anglican Church Grammar School)
11	Oxley Creek
12	Brisbane River
13	Indooroopilly Shoppingtown
14	"The Gabba" (Brisbane Cricket Ground)
15	South Bank arts and recreation precinct
16	Central business district
17	University of Queensland (UQ) St Lucia Campus
18	City Botanic Gardens
19	Queensland University of Technology (QUT) Gardens Point Campus
20	Goodwill Bridge (pedestrians and pedal cycles)
21	The Royal Brisbane and Women's Hospital and the Royal Children's Hospital

22	Mater Private Hospital

23	Roma Street Rail Station

24	Roma Street Parkland

25	New Farm Park and Powerhouse

26	Victoria Park Golf Course

27	Brisbane Exhibition Ground

28	Brisbane Riverwalk (Destroyed in 2011 floods)

29	Inner City Bypass (rail) (left) (road) (right)

30	Indooroopilly Golf Course (Long Pocket)

Climate

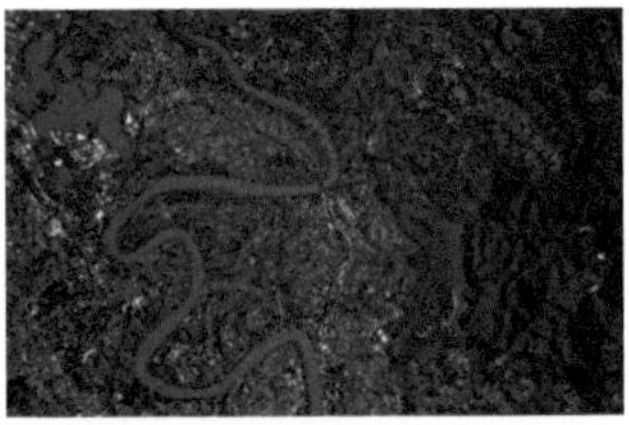
This detailed astronaut photograph illustrates flooding in suburbs of the Brisbane metropolitan region during early 2011.

Brisbane has a humid subtropical climate (Köppen climate classification *Cfa*) with warm to hot and humid summers and dry, moderately warm winters.[44] [45] From November to March, thunderstorms are common over Brisbane, with the more severe events accompanied by large damaging hail stones, torrential rain and destructive winds.

The city's highest recorded temperature was 43.2 °C (110 °F) on 26 January 1940. On 19 July 2007, Brisbane's temperature fell below the freezing point for the first time since records began, registering –0.1 °C (31.8 °F) at the airport.[46] In 2009 Brisbane recorded its hottest winter day at 35.4 °C (95.7 °F) on 24 August.[47] Brisbane's wettest day occurred on January 21, 1887, when 465 millimetres (18.3 in) of rain fell on the city, the highest maximum daily rainfall of Australia's capital cities.

From 2001 until 2010, Brisbane and surrounding temperate areas had been experiencing the most severe drought in over a century, with dam levels dropping to 16.9% of their capacity on 10 August 2007. Residents were mandated by local laws to observe level 6 water restrictions on gardening and other outdoor water usage. Per capita water usage is below 140 litres per day, giving Brisbane one of the lowest per capita usages of water of any Western city in the world.[48] On Sunday January 9, 2011, a weak cyclone crossed north of Brisbane and dropped rainfull on the already saturated South East Coast of Qld, resulting in severe flooding and damage in Brisbane and the surrounding area,[49] ironically the same storm season also resulted in the water storage climbing to over 98% of maximum capacity and breaking the drought.[50] Water restrictions have been replaced with water conservation measures that aim at a target of 200 litres per day/per person, but consumption is rarely over 160 litres. Dust storms in Brisbane are extremely rare; on 23 September 2009, however, a severe dust storm blanketed Brisbane, as well as other parts of eastern Australia.[51] [52]

Brisbane also lies in the Tropical Cyclone risk area. Although cyclones hitting Brisbane are rare, they have happened in the past. The last cyclone to affect Brisbane but not directly cross the city was Tropical Cyclone Hamish in March 2009, although the cyclone remained approx 350 km (220 mi) north of Brisbane[53] but caused significant damage to beaches and caused the worst oil spill in Moreton Bay.[54] Average annual temperature of the sea is 24 °C (75 °F), from 21 °C (70 °F) in July to 27 °C (81 °F) in February.[55]

In November 2011, Brisbane saw 22 days with no recorded rainfall, which is the driest start to November since 1919.[56]

Brisbane City Hall home to the Museum of Brisbane and Brisbane City Council offices.

Governance

Parliament House is where Queensland's Parliament meets.

Unlike other Australian capital cities, a large portion of the greater metropolitan area of Brisbane is controlled by a single local government entity, the Brisbane City Council, inside the City of Brisbane LGA. Since the creation of the Brisbane City Council in 1925 the urban areas of Brisbane have expanded considerably past the City Council boundaries.[57] Prior to that, a far smaller area (comprising the inner suburbs of Brisbane today) was controlled by the Brisbane Municipal Council.

The City is the largest local government body (in terms of population and budget) in Australia. It was formed by the merger of twenty smaller LGAs in 1925, and covers an area of 1367 km^2 (528 sq mi). The Council's annual budget is approximately AUD 1.6 billion, and it has an asset base of AUD 13 billion.[58]

The remainder of the metropolitan area falls into the LGAs of Logan City to the south, Moreton Bay Region in the northern suburbs, the City of Ipswich to the south west, Redland City to the south east on the bayside, with a small strip to the far west of the Scenic Rim Region.

Economy

Brisbane has the largest economy of any city between Sydney and Singapore, which includes cities like Jakarta, Darwin, Townsville and Cairns, and has seen consistent economic growth in recent years as a result of the resources boom. White-collar industries include information technology, financial services, higher and public sector administration generally concentrated in and around the central business district and recently established office areas in the inner suburbs. Blue-collar industries, including petroleum refining, stevedoring, paper milling, metalworking and QR railway workshops, tend to be located on the lower reaches of the Brisbane River and in new industrial zones on the urban fringe. Tourism is an important part of the Brisbane economy, both in its own right and as a gateway to other areas of Queensland.[59]

Skyscrapers in Brisbane's CBD

Since the late 1990s and early 2000s, the Queensland State Government has been developing technology and science industries in Queensland as a whole, and Brisbane in particular, as part of its "Smart State" initiative.[60] The government has invested in several biotechnology and research facilities at several universities in Brisbane. The Institute for Molecular Bioscience at the University of Queensland (UQ) Saint Lucia Campus is a large CSIRO and Queensland state government initiative for research and innovation that is currently being emulated at the Queensland University of Technology (QUT) Campus at Kelvin Grove with the establishment of the Institute of Health and Biomedical Innovation (IHBI).[61]

Brisbane is one of the major business hubs in Australia.[62] Most major Australian companies, as well as numerous international companies, have contact offices in Brisbane, while numerous electronics businesses have distribution hubs in and around the city. DHL Global's Oceanic distribution warehouse is located in Brisbane, as is Asia Pacific Aerospace's headquarters. Home grown major companies include Suncorp-Metway Limited, Flight Centre, Sunsuper, Orrcon, Credit Union Australia, Boeing Australia, Donut King, Wotif.com, WebCentral, PIPE Networks, Krome Studios, NetBox Blue, Mincom Limited, TechnologyOne and Virgin Australia. Brisbane has the fourth highest median household income of the Australian capital cities at AUD 57,772.[63]

Port of Brisbane

The Port of Brisbane is on the lower reaches of the Brisbane River and on Fisherman's Island at the rivers mouth, and is the 3rd most important port in Australia for value of goods.[64] Container freight, sugar, grain, coal and bulk liquids are the major exports. Most of the port facilities are less than three decades old and some are built on reclaimed mangroves and wetlands.

Port of Brisbane.

The Port is a part of the Australia TradeCoast, the country's fastest-growing economic development area.[65] Geographically, Australia TradeCoast occupies a large swathe of land around the airport and port. Commercially, the area has attracted a mix of companies from throughout the Asia Pacific region.[65]

Retail

Queen Street Mall.

Brisbane has a range of retail precincts, both in the Central Business District and in surrounding suburbs. The Queen Street Mall has a vast array of cafes, restaurants, cinemas, gift shops and shopping centres including: Wintergarden, Broadway on the Mall, QueensPlaza, Brisbane Arcade, Queen Adelaide Building, Tattersalls Arcade and The Myer Centre. Edward St, in the city, is home to many luxury international retailers, such as MaxMara, Gucci, Bally, Prada, Canali, Mont Blanc, Louis Vuitton, Tiffany and Co., Oroton, Polo Ralph Lauren, Bvlgari, Hugo Boss, Hermes, and more. The majority of retail business is done within the suburbs of Brisbane in shopping centres which include major department store chains. There are 3 major Westfield shopping centres in Brisbane located in the suburbs of Chermside (Westfield Chermside), Mount Gravatt (Westfield Garden City) and Carindale (Westfield Carindale).[66]

Other large shopping centres exist at Indooroopilly (Indooroopilly Shopping Centre), Toombul (Centro Toombul) and Mitchelton (Brookside Shopping Centre). Other major shopping centres through-out the metropolitan area include North Lakes (Westfield North Lakes), Strathpine (Westfield Strathpine) and Loganholme (Logan Hyperdome).

Demographics

Chinatown, Brisbane.

Significant overseas born populations[67]

Country of birth	Population (2006)
United Kingdom	95,315
New Zealand	73,128
South Africa	12,824
Vietnam	11,857
China	11,418
Spain	10,880
Philippines	9,871
Germany	8,645
India	7,549
Netherlands	7,014

Fiji	6,791
Papua New Guinea	6,706
Italy	6,704
Malaysia	6,686
United States	6,057
Hong Kong	6,036
South Korea	4,841
Sri Lanka	4,806

Brisbane population by year		
1859	6,000	
1942	750,000	
2010	2,043,185[68]	Estimate
2026	2,681,100[68]	Projected
2056	3,979,300[68]	Projected

The statistical division of Brisbane includes much of Brisbane's Local Government Area as well as the cities of Ipswich, Moreton Bay, Logan City and Redland City which demographically are part of a single conurbation. The 2006 census reported 1,763,131 residents within the Brisbane Statistical Division, making it the third largest city in Australia.[69] Brisbane recorded the largest growth rate of all capital cities in the last Census, with an annual growth rate of 2.2%.[70] The median age across the city was 35 years.[3]

The 2006 census showed that 1.7% of Brisbane's population were of indigenous origin and 21.7% were born overseas. Of those born outside of Australia, the three main countries of birth were New Zealand, South Africa, and the United Kingdom.

Approximately 16.1% of households spoke a language other than English, with the most common languages being Mandarin 1.1%, Vietnamese 0.9%, Cantonese 0.9%, Italian 0.6% and Samoan 0.5%. Areas of significant overseas populations include the southern suburbs of Brisbane. Moorooka is home to many residents of African descent. Most of the Vietnamese population reside in the suburbs of Darra and Inala. Sunnybank is where the majority of Brisbane's Chinese population reside, consisting mainly of people from Republic of China (Taiwan) and Hong Kong, while those from China, Malaysia and other Southeast Asian countries are often found not in one particular area but all around Brisbane, especially in the southern suburbs. Brisbane has the highest population of Taiwanese Australian citizens. It has been estimated that the population has grown to an estimated 35 000+, making them the highest Asian population in Brisbane. Consequently, Sunnybank and its surrounding suburbs have often been dubbed as the "Real Chinatown" and "Taiwan Town".

The inner southern suburbs were considered the most densely populated areas of people with Southern European heritage, primarily Greece and Italy, as well as Cyprus and other nations in Southeastern Europe. New Farm was historically home to many of Brisbane's early Italian community. There are also a major number of Indians, Filipinos, Dutch, Hispanics, Germans, Koreans, Papua New Guineans, Fijians and other Pacific Islander communities in the city.

Education

The Forgan Smith Building and the Great Court, University of Queensland.

Brisbane has multi-campus universities and colleges including Griffith University, Queensland University of Technology and the University of Queensland. Other universities which have campuses in Brisbane include the Australian Catholic University, Central Queensland University, James Cook University, JMC Academy, University of Southern Queensland and the University of the Sunshine Coast.

There are three major TAFE colleges in Brisbane; the Brisbane North Institute of TAFE, the Metropolitan South Institute of TAFE, and the Southbank Institute of TAFE.[71] Brisbane is also home to numerous other independent tertiary providers, including the Australian College of Natural Medicine, the Brisbane College of Theology, QANTM, as well as Jschool: Journalism Education & Training.

The majority of Brisbane's preschool, primary, and secondary schools are run under the jurisdiction of Education Queensland, a branch of the Queensland Government.[72] There are also a large number of independent and Roman Catholic run schools.

Culture

Arts and entertainment

Brisbane has a growing live music scene, both popular and classical. The Queensland Performing Arts Centre (QPAC), which is located at South Bank, consists of the Lyric Theatre, a Concert Hall, Cremorne Theatre and the Playhouse Theatre. The Queensland Ballet, Opera Queensland, Queensland Theatre Company and other performance art groups stage performances in the different venues. It is also the major performing venue for The Queensland Symphony Orchestra, Brisbane's only professional symphony orchestra and Queensland's largest performing arts company. The Queensland Conservatorium, in which professional companies and Conservatorium students also stage performances, is located within the South Bank Parklands.

Numerous choirs present performances across the city annually. These choirs include the Brisbane Chorale, Queensland Choir, Brisbane Chamber Choir, Canticum Chamber Choir, Brisbane Concert Choir, Imogen Children's Chorale and Brisbane Birralee Voices. Due to the lack of a suitable purpose built performance venue for choral music, these choirs typically perform in the city's many churches. The Queensland Gallery of Modern Art (GOMA), opened in December 2006, is one of the latest additions to the South Bank precinct and houses some of the most well-known pieces of modern art from within and outside Australia. GOMA is the largest modern art gallery in Australia. GOMA holds the Asia Pacific Triennial (APT) which focuses on contemporary art from the Asia and Pacific in a variety of media from painting to video work. In Addition, its size enables the gallery to exhibit particularly large shows — the Andy Warhol exhibition being the largest survey of his work in Australia. GOMA also boasts Australia's largest purpose-built Cinémathèque. The Gallery of Modern Art is located next to the State Library of Queensland and the Queensland Art Gallery. Along with Beijing, Berlin, Birmingham and Marseille, Brisbane was nominated as one of the Top 5 International Music Hotspots by Billboard in 2007. There are also popular entertainment pubs and clubs within both the City and Fortitude Valley.[73] [74]

The Brisbane Powerhouse in New Farm and the Judith Wright Centre of Contemporary Arts on Brunswick Street in Fortitude Valley also feature diverse programs featuring exhibitions and festivals of visual art, music and dance. The La Boite Theatre Company performs at the Roundhouse Theatre at Kelvin Grove. Twelfth Night Theatre at Bowen Hills is also a professional theatre. The Powerhouse complex stages a range of productions. There are numerous amateur theatre groups in Brisbane. The oldest is the Brisbane Arts Theatre which was founded in 1936. It has a regular adult and children's theatre and is located in Petrie Terrace.

Annual events

Fireworks over the South Bank Parklands

Major cultural events in Brisbane include the Ekka (the Royal Queensland Show), held each August, and the Riverfestival, held each September at South Bank Parklands and surrounding areas. Warana, (meaning *Blue Skies*), was a former spring festival which began in 1961 and was held in September each year. Run as a celebration of Brisbane, Warana was similar to Melbourne's Moomba festival. In 1996 the annual festival was changed to a biennial Brisbane Festival.[75]

The Brisbane International Film Festival (BIFF) is held in July/August in a variety of venues around Brisbane including the Regent Cinema in Queen Street Mall. Brisbane's historic Regent Cinema, on the Queen Street Mall, closed on Monday 14 June 2010. BIFF features new films and retrospectives by domestic and international filmmakers along with seminars and awards.

The Paniyiri festival at Musgrave Park (corner of Russell and Edmondstone Streets, South Brisbane) is an annual Greek cultural festival held on the first weekend in May. The Brisbane Medieval Fayre and Tournament is held each June in Musgrave Park.

The Valley Fiesta is an annual three-day event organised by the Valley Chamber of Commerce. It was launched by Brisbane Marketing in 2002 to promote Fortitude Valley as a hub for arts and youth culture. It features free live music, market stalls, food and drink from many local restaurants and cafés, and other entertainment.

The Bridge to Brisbane fun run has become a major annual charity event for Brisbane.

Tourism and recreation

Sir Thomas Brisbane Planetarium at the base of Mount Coot-tha.

Tourism plays a major role in Brisbane's economy, being the third-most popular destination for international tourist after Sydney and Melbourne.[76] Popular tourist and recreation areas in Brisbane include the South Bank Parklands, Roma Street Parkland, the City Botanic Gardens, Brisbane Forest Park and Portside Wharf. The Lone Pine Koala Sanctuary opened in 1927 and was the world's first koala sanctuary.[77]

The suburb of Mount Coot-tha is home to a popular state forest, and the Brisbane Botanic Gardens which houses the Sir Thomas Brisbane Planetarium and the "Tsuki-yama-chisen" Japanese Garden (formerly of the Japanese Government Pavilion of Brisbane's World Expo '88).

Brisbane has over 27 km (16.8 mi) of bicycle pathways, mostly surrounding the Brisbane river and city centre, extending to the west of the city. The river itself was popular with bathers, and it permitted boating excursions to Moreton Bay when the main port was in the city reaches.[77] Today fishing and boating are more common. Other popular recreation activities include the Story Bridge adventure climb and rock climbing at the Kangaroo Point cliffs.

Sport

An NRL game at Suncorp Stadium.

Brisbane has hosted several major sporting events including the 1982 Commonwealth Games and the 2001 Goodwill Games. The city also hosted events during the 1987 Rugby World Cup, 1992 Cricket World Cup, 2000 Sydney Olympics, the 2003 Rugby World Cup and hosted the Final of the 2008 Rugby League World Cup and will host along with the Gold Coast, some events for the 2018 Commonwealth Games. In 2005, then Premier Peter Beattie announced plans for Brisbane to bid to host the 2024 Olympic Games,[78] which in August 2008 received in principle Australian Olympic Committee support, including that of the Queensland Premier Anna Bligh and Brisbane Lord Mayor Campbell Newman.[79] Brisbane is also home to a professional football (soccer) team named Brisbane Roar Football Club. They are in the Hyundai A-League.

Queensland Tennis Centre at Brisbane International is a professional tennis tournament.

The city's major sporting venues include Brisbane Cricket Ground, Sleeman Centre at Chandler, Suncorp Stadium, Ballymore Stadium and the stadium facilities of the Queensland Sport and Athletics Centre in Nathan. With the closure of the Milton Tennis grounds in 1994, Brisbane lacked a major tennis facility. In 2005, the State Government approved the State Tennis Centre a new A$65 million tennis stadium. The construction was completed in 2008. The Brisbane International is held here from January 2009.

Brisbane has teams in all major interstate competitions, excluding the National Basketball League.

Sport	Team Name	League	Stadium	Reference
Rugby League	Queensland	State of Origin	Suncorp Stadium	[80]
	Brisbane Broncos	National Rugby League		[81]
Rugby Union	Queensland Reds	Super Rugby		[82]
Football	Brisbane Roar	A-League		[83]
Cricket	Queensland Bulls	Sheffield Shield Ryobi One Day Cup KFC Twenty20 Big Bash	The Gabba	[84]
Australian rules football	Brisbane Lions	Australian Football League		[85]
Netball	Queensland Firebirds	ANZ Championship	Brisbane Convention and Exhibition Centre	[86]
Baseball	Brisbane Bandits	Australian Baseball League	Brisbane Exhibition Ground	

Media

The main newspapers of Brisbane are *The Courier-Mail* and *The Sunday Mail*, both owned by News Corporation. Brisbane receives the national daily, *The Australian*, and the *Weekend Australian*, together with Fairfax papers *Australian Financial Review*, the *Sydney Morning Herald* and *The Age*, and Fairfax website Brisbane Times. There are community and suburban newspapers throughout the metropolitan and regional areas, including *Brisbane News* and *City News*, many of which are produced by Quest Community Newspapers. *mX*, a free daily commuter newspaper, was launched in 2007, following the newspaper's success in Melbourne and Sydney.

Brisbane is served by all five major television networks in Australia, which broadcast from the summit of Mount Coot-tha. The three commercial stations, Seven, Nine, and Ten, are accompanied by two government networks, ABC and SBS, with all five providing digital television. New digital-only channels available in addition to ABC1, Seven, Nine, Ten and SBS One include One HD, Eleven, ABC2, ABC3, ABC News 24, SBS Two, 7Two, 7mate, GEM HD and GO!. 31, a community station, also broadcasts in Brisbane. Optus, Foxtel and Austar all operate PayTV services in Brisbane, via cable and satellite means.

The ABC transmits all five of its radio networks to Brisbane; 612 ABC Brisbane, ABC Classic FM, ABC NewsRadio, Radio National, and Triple J. SBS broadcasts its national radio network. Brisbane is serviced by major commercial radio stations, including 4KQ, 4BC, 4BH, 97.3 FM, B105 FM, Nova 106.9, and Triple M. Brisbane is also serviced by major community radio stations such as 96five Family FM, 4MBS Classic FM 103.7, 4EB FM and 4ZZZ 102.1.

Infrastructure

Health

Brisbane is covered by Queensland Health's "Northside" and "Southside" health service areas.[87] Within the greater Brisbane area there are 8 major public hospitals, 4 major private hospitals, and smaller public and private facilities. Specialist and general medical practices are located in the CBD, and most suburbs and localities.

Aerial of part of Mater Health Services campus at South Brisbane.

Transport

Brisbane has an extensive transportation network within the city, as well as connections to regional centres, interstate and to overseas destinations. The use of urban public transport is still only a small component of total passenger transport, the largest component being travel by private car.[88]

CityCat on the Brisbane River.

Public transport is provided by bus, rail and ferry services. Bus services are operated by public and private operators whereas trains and ferries are operated by public agencies. The Brisbane central business district (CBD) is the central hub for all public transport services with services focusing on Queen Street Bus Station, Roma Street and Central railway stations, and various city ferries wharves. Brisbane's CityCat high speed ferry service, popular with tourists and commuters, operates services along the Brisbane River between the University of Queensland and Apollo Road.

The Citytrain urban rail network consists of 10 suburban lines and covers mostly the west, north and east sides of the city. It also provides the route for an Airtrain service under joint public/private control between the City and Brisbane Airport. Since 2000, Brisbane has been developing a network of busways, including the South East Busway and the Inner Northern Busway, to provide faster bus services. TransLink operates an integrated ticketing system across the public transport network.

CityTrain Suburban Multiple Unit at Nambour Station.

The Brisbane River has created a barrier to some road transport routes. In total there are ten road bridges, mostly concentrated in the inner city area. This has intensified the need for transport routes to focus on the inner city. There are also three railway bridges and two pedestrian bridges. The Eleanor Schonell Bridge (originally named, and still generally known as, The Green Bridge) between the University of Queensland and Dutton Park is for use by buses, pedestrians and cyclists. There are currently multiple tunnel and bridge projects underway as part of the TransApex plan.

Houghton Highway, the longest bridge in Australia.

An extensive network of pedestrian and cyclist pathways have been created along the banks of the Brisbane River to form a Riverwalk network.[89]

International Terminal at Brisbane Airport.

Brisbane is served by several urban and inter-urban motorways. The Pacific Motorway connects the central city with the Gold Coast to the south. The Ipswich Motorway connects the city with Ipswich to the west via the southern suburbs, while the Western Freeway and the Centenary Motorway provide a connection between Brisbane's inner-west and the outer south-west, connecting with the Ipswich Motorway south of the Brisbane River. The Bruce Highway is Brisbane's main route north of the city to the rest of the State. The Bruce Highway terminates 1700 km (1056 mi) away in Cairns and passes through most major cities along the Queensland coast. The Gateway Motorway is a private toll road which connects the Gold Coast and Sunshine Coasts by providing an alternate route via the Gateway Bridge avoiding Brisbane's inner city area. The Port of Brisbane Motorway links the Gateway to the Port of Brisbane, while Inner City Bypass and the Riverside Expressway act as the inner ring freeway system to prevent motorists from travelling through the city's congested centre.[90]

Brisbane's population growth placed strains on South East Queensland's transport system. The State Government and Brisbane City Council have responded with infrastructure plans and increased funding for transportation projects, such as the South East Queensland Infrastructure Plan and Program. Most of the focus has been placed on expanding current road infrastructure, particularly tunnels and bypasses, as well as improving the public transport system.

Brisbane Airport (IATA code: *BNE*) is the city's main airport, the third busiest in Australia after Sydney Airport and Melbourne Airport. It is located north-east of the city centre and provides domestic and international passenger services. In the 2008–2009 year, Brisbane Airport handled over 18.5 million passengers. The airport is serviced by the Brisbane Airtrain which provides a rail service from Brisbane's city centre to and from the airport. Archerfield Airport (in Brisbane's southern suburbs) acts as a general aviation airport.

King George Square Busway Station, an underground bus station.

Utilities

Lake Wivenhoe, Brisbane's primary water reserve.

Water storage, treatment and delivery for Brisbane is handled by SEQ Water, which sells on to Queensland Urban Utilities (previously Brisbane Water) for distribution to the greater Brisbane area. Water for the area is stored in one of three dams; Wivenhoe, Somerset and North Pine. Since 13 May 2005, Brisbane has enforced water restrictions due to drought.[91] This has also led to the State Government announcing that purified recycled water would be pumped into the dams once the pipeline was complete in 2009.[92] More recently, restrictions were lifted and water conservation is no longer a major concern of council, although residents are asked to follow permanent conservation measures.

Electricity and gas grids in Brisbane are handled by Energex (electricity), and Origin Energy (gas), with each company previously holding a monopoly on domestic retail supply. Since 1 July 2007 Queensland regulation changes have opened up the retail energy market, allowing multiple companies to resell both gas and electricity.[93]

Metropolitan Brisbane is serviced by all major and most minor telecommunications companies and their networks. Brisbane has the largest number of enabled DSL telephone exchanges in Queensland. An increasing number are also enabled with special hardware (DSLAMs) which enable high speed ADSL2+ internet access. The Brisbane CBD also features a complete underground fibre optics network, with numerous connections to the inner suburbs provided by various service providers.

Telstra and Optus provide both high speed internet as well as Pay TV through their cable services for the bulk of the city's metropolitan area. Both of these providers also host wireless networks with hotspots within both the inner and suburban areas. In addition, 3 Mobile, Telstra, Optus and Vodafone all operate both 2.5G, 3G and 3.5G mobile phone networks citywide.[94]

In popular culture

Australian Navy references

Three ships and a naval base of the Royal Australian Navy named HMAS Brisbane are named after the city.

Music references

Brisbane is featured in music including The Saints "Brisbane (Security City)" (1978); "Love You Brisbane" theme single from the 1980s; "Footpaths Of Brunswick Street" – The Natives Of Bedlam (1993); Powderfinger's album "Vulture Street" (2003); R.E.M.'s "Parakeet" (from "Up") (1999); The Stranglers "Nuclear Device" (1979); The Go-Betweens' first single "Lee Remick" (1978); Evil Eddie's "Queensland" (2010); and John Kennedy's "Brisvegas"

(2007).

Bris Vegas

"Bris Vegas" is a nickname given to the city. This has been attributed to an Elvis Presley tribute CD[95] and the city's growing live music scene.[96] It is believed to have been first used in print in a 1996 edition of The Courier-Mail,[95] also approximately the time of the opening of the Treasury Casino in Brisbane and the popularisation of poker machines in Brisbane bars and clubs, a play on the popular gaming ground of Las Vegas. The name has also been attributed to the city's nightlife,[97] compact size of the central business district and previous lack of sophistication when compared with more populated Australian cities[98] [99] and also to Las Vegas.[100]

In Television

Brisbane is the setting for the Australian TV series "Slide".

See also

- South East Queensland

Gallery

Brisbane Skyline at night.

View of Brisbane River from Victoria Bridge.

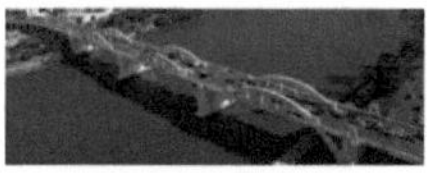
William Jolly Bridge.

View of Brisbane from Victoria Bridge.

Kurilpa Bridge at night, April 2010

References

[1] "3218.0 – Regional Population Growth, Australia, 2009–10" (http://www.abs.gov.au/ausstats/abs@.nsf/Products/3218.0~2009-10~Main+Features~Main+Features?OpenDocument). Australian Bureau of Statistics. 31 March 2011. . Retrieved 31 March 2011.

[2] Australian Bureau of Statistics (17 March 2008). "Explore Your City Through the 2006 Census Social Atlas Series" (http://abs.gov.au/websitedbs/d3310114.nsf/4a256353001af3ed4b2562bb00121564/45b3371f4a681356ca25740e007c92bf!OpenDocument). . Retrieved 19 May 2008.

[3] Australian Bureau of Statistics (25 October 2007). "Community Profile Series : Brisbane (Statistical Division)" (http://www.censusdata.abs.gov.au/ABSNavigation/prenav/ProductSelect?newproducttype=Community+Profiles&collection=Census&period=2006&areacode=305&breadcrumb=LP&currentaction=201&action=401). *2006 Census of Population and Housing*. . Retrieved 21 January 2008. Map (http://www.censusdata.abs.gov.au/ABSNavigation/ImageServer?id=map,Census,2006,305)

[4] *Macquarie ABC Dictionary*. The Macquarie Library Pty Ltd. 2003. p. 121. ISBN 0 876429 37 2.

[5] (http://www.abs.gov.au/ausstats/abs@.nsf/Products/3218.0~2008-09~Main+Features~Queensland?OpenDocument#PARALINK5) Preliminary population estimates for local government areas, 2009. Local government areas included in the calculation of the South East

Queensland population are: Shire of Beaudesert, Shire of Boonah, City of Brisbane, Shire of Caboolture, City of Caloundra, Shire of Esk, Shire of Gatton, Gold Coast City, City of Ipswich, Shire of Kilcoy, Shire of Laidley, Logan City, Shire of Maroochy, Shire of Noosa, Shire of Pine Rivers, Redcliffe City, and Redland City.

[6] "Brisbanian entry in Wiktionary" (http://en.wiktionary.org/wiki/Brisbanian). Wiktionary. . Retrieved 2011-12-09.

[7] "The World According to GaWC 2010" (http://www.lboro.ac.uk/gawc/world2010.html). *Globalization and World Cities (GaWC) Study Group and Network*. Loughborough University. . Retrieved 2011-09-15.

[8] "Tom Petrie's Early Reminiscences of Early Queensland" (http://www.seqhistory.com/index.php?option=com_content&view=article&id=95:part1chpt1&catid=42:tom-petrie&Itemid=67&limitstart=4). . Retrieved 24 November 2008.

[9] Our Brisbane – Our shared vision (http://www.brisbane.qld.gov.au/documents/about council/vision2026_final_ourbrisbane.pdf) – Brisbane City Council Page 2

[10] "Redcliffe" (http://www.smh.com.au/articles/2005/02/17/1108500203689.html). *The Sydney Morning Herald*. 8 February 2004. . Retrieved 17 May 2008.

[11] "John Oxley Governor Report" (http://www.seqhistory.com/john-oxley/139-john-oxley-1823-governorreport?start=2). . Retrieved 1 February 2010.

[12] Potter, Ron. "Place Names of South East Queensland" (http://www.dovenetq.net.au/~piula/Placenames/page55.html). Piula Publications. . Retrieved 17 May 2008.

[13] compiled by Royal Automobile Club of Queensland. (1980). *Seeing South-East Queensland* (2 ed.). RACQ. p. 7. ISBN 0-909518-07-6.

[14] "About Redcliffe" (http://www.redcliffe.qld.gov.au/about_us.htm). Redcliffe City Council. . Retrieved 1 December 2007.

[15] Lybaek, Lena; Konrad Raiser, Stefanie Schardien (2004). *Gemeinschaft der Kirchen und gesellschaftliche Verantwortung*. Münster: LIT. p. 114. ISBN 978-3825870614.

[16] "Christopher Eipper (1813–1894)" (http://web.archive.org/web/20071118093456/http://www.pelicanwaters.com/pelicanwaters-streetsigns.php). *Street Signs —And What They Mean*. Pelican Waters Shire Council. Archived from the original (http://www.pelicanwaters.com/pelicanwaters-streetsigns.php) on 18 November 2007. . Retrieved 20 December 2007.

[17] de Strzelecki, Paul Edmond (1845). *Physical Description of New South Wales and Van Diemen's Land: Accompanied by a Geological Map, Sections, and Diagrams*. London, United Kingdom: Longman, Brown, Green, and Longmans.

[18] Establishing Queensland's borders (http://www.nrw.qld.gov.au/museum/articles_complete/surveying/border.html)

[19] "Organisation chart" (http://web.archive.org/web/20080522132144/http://www.brisbane.qld.gov.au/BCC:STANDARD:827799619:pc=PC_95). Brisbane City Council. Archived from the original (http://www.brisbane.qld.gov.au/BCC:STANDARD:827799619:pc=PC_95) on 22 May 2008. . Retrieved 20 December 2007.

[20] "Jolly, William Alfred (1881–1955)" (http://www.adb.online.anu.edu.au/biogs/A090501b.htm). Australian Dictionary of Biography. . Retrieved 20 December 2007.

[21] "Brisbane" (http://www.anzacday.org.au/education/tff/memorials/queensland.html). ANZAC Day Commemoration Committee (Qld) Incorporated. 1998. . Retrieved 28 December 2007.

[22] http://en.wikipedia.org/wiki/Story_Bridge

[23] Peter Dunn (2 March 2005). "Hirings Section" (http://home.st.net.au/~dunn/ausarmy/hiringsno1lofc.htm). Australia @ War. . Retrieved 7 January 2008.

[24] "QM Supply in the Pacific during WWII" (http://web.archive.org/web/20080520050058/http://www.quartermaster.army.mil/OQMG/professional_bulletin/1999/spring1999/QM+Supply+in+the+Pacific+During+WWII.htm). Quartermaster Professional Bulletin. Spring 1999. Archived from the original (http://www.quartermaster.army.mil/OQMG/professional_bulletin/1999/spring1999/QM Supply in the Pacific During WWII.htm) on 20 May 2008. . Retrieved 7 January 2008.

[25] Peter Dunn (27 August 2005). "The Battle Of Brisbane — 26 & 27 November 1942" (http://home.st.net.au/~dunn/ozatwar/bob.htm). Australia @ War. . Retrieved 7 January 2008.

[26] Brisbane's last in but best-dressed (http://city-news.whereilive.com.au/lifestyle/story/brisbanes-last-in-but-best-dressed/), Brooke Falvey, City news, 11 July 2008.

[27] http://www.couriermail.com.au/ipad/filmmakers-to-honour-brisbane-bard/story-fn6ck51p-1225969834655

[28] She picked me up at a dance one night (http://www.slq.qld.gov.au/whats-on/exhibit/online/travelling/bentson), Joan and Bill Bentson, Queensland Government.

[29] "ACGA Past Games 1982" (http://www.commonwealthgames.org.au/Templates/Games_PastGames_1982.htm). Commonwealth Games Australia. . Retrieved 28 December 2007.

[30] Rebecca Bell. "Expo 88 / Brisbane" (http://www.ozbird.com/oz/OzCulture/expo88/brisbane/default.htm). OZ Culture. . Retrieved 28 December 2007.

[31] Berry, Petrina (13 January 2011). "Brisbane braces for flood peak as Queensland's flood crisis continues" (http://www.couriermail.com.au/news/breaking-news/brisbane-braces-for-flood-peak-as-queenslands-flood/story-fn7ik8u2-1225986784487). *Courier Mail*. . Retrieved 14 January 2011.

[32] Before and after photos of the floods in Brisbane (http://www.abc.net.au/news/infographics/qld-floods/beforeafter2.htm/)

[33] Gregory, Helen (2007). *Brisbane Then and Now*. Wingfield, South Australia: Salamander Books. p. 60. ISBN 9781741730111.

[34] Flood-proof road destroyed in deluge (http://www.abc.net.au/news/stories/2010/10/12/3036241.htm?site=Brisbane) – ABC News (Australia)

[35] Gunn, Angus M. (1978). *Habitat: Human Settlements in an Urban Age*. Pergamon Press. p. 178. ISBN 0080214878.

[36] "Timeline for Brisbane River" (http://www.coastal.crc.org.au/pdf/HistoricalCoastlines/App_3_Timeline_BrisbaneRiver.pdf) (PDF). Coastal CRC. . Retrieved 4 January 2008.

[37] Campbell Newman, *"bmag"*, 3 November 2009

[38] "TimeWalks Brisbane — Windmill" (http://www.slq.qld.gov.au/oh/treasures/timewalks/bris/1870/windmill). Queensland Government. 24 March 2008. . Retrieved 10 April 2008.

[39] Statham-Drew, Pamela (1990). *The Origin of Australia's Capital Cities*. Cambridge University Press. p. 257. ISBN 978-0521408325.

[40] Pike, Jeffrey (2002). *Australia*. Insight. ISBN 978-9812347992.

[41] "The Commissariat Stores" (http://www.queenslandhistory.org.au/comm.html). . Retrieved 24 February 2008.

[42] Population Growth (http://www.abs.gov.au/Ausstats/abs@.nsf/0/83DDE980E589FB5FCA256889000F8EBD?Open) *Australian Bureau of Statistics* – Accessed 28 December 2007

[43] http://buildingdb.ctbuh.org/?do=create>

[44] Tapper, Andrew; Tapper, Nigel (1996). Gray, Kathleen. ed. *The weather and climate of Australia and New Zealand* (First ed.). Melbourne, Australia: Oxford University Press. p. 300. ISBN 0195533933.

[45] Linacre, Edward; Geerts, Bart (1997). *Climates and Weather Explained* (http://books.google.com/?id=mkZa1KLHCAQC&lpg=PA379&pg=PA379#v=onepage&q=). London: Routledge. p. 379. ISBN 0-415-12519-7. .

[46] Daniel Sankey and Tony Moore (19 July 2007). "Coldest day on record for Brisbane" (http://www.brisbanetimes.com.au/articles/2007/07/19/1184559902397.html). The Brisbane Times. . Retrieved 5 January 2008.

[47] Unknown (24 August 2009). "Hot August day as Records Fall" (http://www.brisbanetimes.com.au/queensland/hot-august-day-as-records-fall-20090824-evxp.html). The Brisbane Times. . Retrieved 31 August 2010.

[48] "Brisbane residents best water savers in world: Newman" (http://www.abc.net.au/news/stories/2007/08/27/2016895.htm). ABC News. . Retrieved 19 March 2008.

[49] . "Raging floods bear down on Brisbane" (http://www.abc.net.au/news/2011-01-11/raging-floods-bear-down-on-brisbane/1901406). .

[50] "SEQWater latest dam levels" (http://seqwater.com.au/public/dam-levels). .

[51] Cubby, Ben (23 September 2009). "Global warning: Sydney dust storm just the beginning" (http://www.brisbanetimes.com.au/opinion/society-and-culture/global-warning-sydney-dust-storm-just-the-beginning-20090923-g1fi.html). *Brisbane Times* (Brisbane). . Retrieved 25 September 2009.

[52] Brisbane on alert as dust storms sweep east (http://www.abc.net.au/news/stories/2009/09/23/2694096.htm)

[53] "Severe Tropical Cyclone Hamish" (http://www.bom.gov.au/announcements/sevwx/qld/qldtc20090304.shtml). Australian Bureau of Meteorology. 2010. . Retrieved 11 August 2010.

[54] Staff Writer (11 March 2009). "Ship spills chemicals in waters off Australia" (http://edition.cnn.com/2009/WORLD/asiapcf/03/11/aus.chemicalspill/). Cable News Network. . Retrieved 11 March 2009.

[55] "Brisbane Climate Guide" (http://www.weather2travel.com/climate-guides/australia/queensland/brisbane.php). . Retrieved October 2011.

[56] "November dry spell in Brisbane set to end as rain forecast" (http://www.couriermail.com.au/news/queensland/time-to-chill-out-as-scorchers-near-end/story-e6freoof-1226201906261). . Retrieved November 2011.

[57] "Brisbane City Council" (http://www.netcat.com.au/NETCAT/STANDARD/PC_4.html). NetCat. . Retrieved 28 December 2007.

[58] "Annual Report and Financial Statements" (http://web.archive.org/web/20070930015350/http://www.brisbane.qld.gov.au/BCC:STANDARD:1658601192:pc=PC_1297). Brisbane City Council. Archived from the original (http://www.brisbane.qld.gov.au/BCC:STANDARD:1658601192:pc=PC_1297) on 30 September 2007. . Retrieved 5 January 2008.

[59] Department of Tourism, Regional Development and Industry (14 December 2007). "Brisbane's business visitors drive $412 million domestic tourism increase" (http://www.brisbanemarketing.com.au/\news-and-events\news-article.aspx?id=171). Brisbane Marketing. . Retrieved 29 December 2007.

[60] "What is the Smart State" (http://www.smartstate.qld.gov.au/strategy/index.shtm#what). Queensland Government. . Retrieved 29 December 2007.

[61] Peter Beattie (4 December 2007). "Brain power drives Smart State" (http://www.news.com.au/couriermail/story/0,23739,22867846-27197,00.html). *The Courier Mail*. News.com.au. . Retrieved 29 December 2007.

[62] "Brisbane business visitor numbers skyrocket" (http://www.etravelblackboard.com/index.asp?id=73027&nav=13). *Brisbane Marketing Convention Bureau*. e-Travel Blackboard. 3 January 2008. . Retrieved 13 January 2008.

[63] "2006 Census QuickStats by Location" (http://www.censusdata.abs.gov.au/ABSNavigation/prenav/PopularAreas?ReadForm&prenavtabname=Popular Locations&type=popular&&navmapdisplayed=true&javascript=true&textversion=false&collection=Census&period=2006&producttype=QuickStats&method=&productlabel=&breadcrumb=PL&topic=&). *Australian Bureau of Statistics*. . Retrieved 19 July 2008.

[64] "Brisbane Container Terminal, Australia" (http://72.14.253.104/search?q=cache:OpXT0QTpjWEJ:www.port-technology.com/projects/brisbane/index.html+http://www.port-technology.com/projects/brisbane/index.html&hl=en&ct=clnk&cd=1&gl=au). Port Technology. . Retrieved 29 December 2007.

[65] "About Us" (http://web.archive.org/web/20080108210134/http://www.australiatradecoast.com.au/AboutAustraliaTradeCoast/index.aspx). Australia TradeCoast. Archived from the original (http://www.australiatradecoast.com.au/AboutAustraliaTradeCoast/index.aspx) on 8 January 2008. . Retrieved 13 January 2008.

[66] "History" (http://www.westfield.com/corporate/about/history.html). Westfield Group. . Retrieved 29 December 2007.

[67] Australian Bureau of Statistics (25 October 2007). "Community Profile Series : Brisbane (Major Statistical Region)" (http://www.censusdata.abs.gov.au/ABSNavigation/prenav/ProductSelect?newproducttype=Community+Profiles&collection=Census&period=2006&areacode=31&breadcrumb=LP&currentaction=201&action=401). *2006 Census of Population and Housing*. . Retrieved 27 December 2009. refer "Basic Community Profile – Brisbane" sheet B10

[68] "Growth in Melbourne" (http://www.abs.gov.au/ausstats/abs@.nsf/Products/3218.0~2007-08~Main+Features~Victoria?OpenDocument). *3218.0 – Regional Population Growth, Australia, 2007–08*. Australian Bureau of Statistics. 23 April 2009. . Retrieved 11 September 2009.

[69] Australian Bureau of Statistics (25 October 2007). "2006 Census QuickStats: Brisbane (Statistical Division)" (http://www.censusdata.abs.gov.au/ABSNavigation/prenav/LocationSearch?collection=Census&period=2006&areacode=305&producttype=QuickStats&breadcrumb=PL&action=401). *2006 Census QuickStats*. . Retrieved 21 March 2008.

[70] "3218.0 – Regional Population Growth, Australia, 1996 to 2006" (http://www.abs.gov.au/AUSSTATS/abs@.nsf/Latestproducts/3218.0Main Features31996 to 2006?opendocument&tabname=Summary&prodno=3218.0&issue=1996 to 2006&num=&view=#CAPITAL CITY GROWTH). Australian Bureau of Statistics. . Retrieved 3 March 2008.

[71] "TAFE Queensland" (http://www.tafe.qld.gov.au/dds/search/browseLocations.do?call_centre_mode=false&externalCallMode=false&breadCrumbsBase=<a+href="/"+title="Home">Home</a>&ins_spec=false). Queensland Government. . Retrieved 2 December 2007.

[72] "Education Queensland" (http://education.qld.gov.au/eq/). Queensland Government. . Retrieved 2 December 2007.

[73] "Billboard Loves Brisbane" (http://www.abc.net.au/triplej/musicnews/s1838651.htm). *Music News*. Triple J. . Retrieved 15 November 2007.

[74] "Beijing, Berlin among music hot spots in 2007" (http://www.reuters.com/article/musicNews/idUSN0126189720070102?pageNumber=2&virtualBrandChannel=0&sp=true). *Music News*. Reuters. 1 January 2007. . Retrieved 29 December 2007.

[75] "History" (http://web.archive.org/web/20080220023411/http://www.brisbanefestival.com.au/history.html). Brisbane Festival. Archived from the original (http://www.brisbanefestival.com.au/history.html) on 20 February 2008. . Retrieved 2 March 2008.

[76] "International Market Tourism Facts" (http://www.tourismaustralia.com/content/Research/Factsheets/TopTen_Regions_Dec2006.pdf) (PDF). Tourism Australia. .

[77] Gregory, Helen (2007). *Brisbane Then and Now*. Wingfield, South Australia: Salamander Books. p. 140. ISBN 9781741730111.

[78] Eleanor Hall (1 April 2005). "Brisbane keen to bid for 2024 Olympics" (http://www.abc.net.au/worldtoday/content/2005/s1336250.htm). *The World Today*. ABC. . Retrieved 5 January 2008.

[79] Brisbane could host Olympics in 2024 (http://news.ninemsn.com.au/article.aspx?id=609048)

[80] "Club Info" (http://www.nrl.com/Clubs/Broncos/tabid/10255/default.aspx). National Rugby League. . Retrieved 30 December 2007.

[81] "Origin of State Colours Queensland Maroons & NSW Blues" (http://www.rl1908.com/Origin/colours.htm). RL1908. . Retrieved 30 December 2007.

[82] "Our History" (http://web.archive.org/web/20080123125252/http://www.queenslandreds.com.au/qru/qru.rugby/page/62650). Queensland Rugby Union. Archived from the original (http://www.queenslandreds.com.au//qru/qru.rugby/page/62650) on 23 January 2008. . Retrieved 30 December 2007.

[83] "History" (http://www.brisbaneroar.com.au/). Brisbane Roar FC. . Retrieved 30 December 2007.

[84] "Introduction" (http://www.qldcricket.com.au/default.asp?PageID=2). Queensland Bulls. . Retrieved 30 December 2007.

[85] "All About the Brisbane Lions" (http://www.lions.com.au/TheClub/History/BrisbaneLions/tabid/5161/Default.aspx). Brisbane Lions. . Retrieved 30 December 2007.

[86] "History of Netball Queensland" (http://www.netballq.org.au/extra.asp?id=78&OrgID=3). Netball Queensland. . Retrieved 30 December 2007.

[87] "Profiles — Hospitals" (http://www.health.qld.gov.au/wwwprofiles/default.asp). Queensland Health. . Retrieved 26 March 2008.

[88] "Year Book Australia, 2005" (http://www.abs.gov.au/Ausstats/abs@.nsf/94713ad445ff1425ca25682000192af2/d81efef6e2252cf4ca256f7200833049!OpenDocument). ABS. . Retrieved 19 February 2008.

[89] "About RiverWalk" (http://web.archive.org/web/20080522142648/http://www.brisbane.qld.gov.au/BCC:STANDARD::pc=PC_1217). Brisbane City Council. Archived from the original (http://www.brisbane.qld.gov.au/BCC:STANDARD::pc=PC_1217) on 22 May 2008. . Retrieved 5 January 2008.

[90] "The upgrade" (http://www.gatewayupgradeproject.com.au/asp/index.asp?sid=5&page=upgradeIntro). Gateway Upgrade Project. . Retrieved 5 January 2008.

[91] Emma Chalmers, Jeremy Pierce and Neil Hickey (8 February 2008). "Queensland Water Commission retain restrictions" (http://www.news.com.au/couriermail/story/0,23739,23178059-952,00.html). *The Courier Mail*. . Retrieved 2 March 2008.

[92] Peter Beattie. "SEQ Will Ave Purified Recycled Water But No Vote: Premier" (http://statements.cabinet.qld.gov.au/MMS/StatementDisplaySingle.aspx?id=50056) (Ministerial media statement). Queensland Government. . Retrieved 29 February 2008.

[93] "Full Retail Competition" (http://web.archive.org/web/20071102094936/http://www.energy.qld.gov.au/frc.cfm). Queensland Department of Mines and Energy. Archived from the original (http://www.energy.qld.gov.au/frc.cfm) on 2 November 2007. . Retrieved 2 March 2008.

[94] Roland Tellzen (1 April 2008). "Mobile broadband takes off" (http://www.australianit.news.com.au/story/0,24897,23460734-5013037,00.html). *The Australian*. news.com.au. . Retrieved 17 April 2008.

[95] Tilston, John. *Meanjin to Brisvegas: Brisbane Comes of Age*. pp. 147–148. ISBN 1-4116-5216-9..

[96] "Billboard Backs Brisvegas" (http://www.theage.com.au/news/music/billboard-backs-brisvegas/2007/01/25/1169594410115.html?page=fullpage#contentSwap1). *The Age*. 25 January 2007. . Retrieved 4 April 2008.

[97] "City Guide: Brisbane" (http://news.bbc.co.uk/sport1/hi/rugby_union/rugby_world_cup/venues_guide/2982769.stm). BBC Sport. 25 September 2003. . Retrieved 4 April 2008.

[98] "QLD: From Brisvegas to Brismanhattan" (http://www.highbeam.com/doc/1P1-98858942.html). AAP General News. 9 September 2004. . Retrieved 4 April 2008.

[99] "Brisbane Residents Embrace City Living" (http://www.theage.com.au/news/National/Brisbane-residents-embrace-city-living/2005/11/02/1130823270850.html). *The Age*. 2 November 2005. . Retrieved 4 April 2008.

[100] Amanda Horswill (15 May 2007). "What's in a Name?" (http://www.news.com.au/couriermail/story/0,23739,21727715-27197,00.html). *The Courier Mail*. . Retrieved 4 April 2008.

External links

- Brisbane travel guide from Wikitravel
- City of Brisbane (http://www.brisbane.qld.gov.au/)
- Brisbane street map (http://www.openstreetmap.org/?lat=-27.4&lon=153&zoom=10)
- Official tourism website of Brisbane (http://www.visitbrisbane.com.au/)

Creative_Commons_licenses

A **Creative Commons license** is one of several copyright licenses that allow the distribution of copyrighted works. The licenses differ by several combinations that condition the terms of distribution. They were initially released on December 16, 2002 by Creative Commons, a U.S. non-profit corporation founded in 2001.

As of July, 2011, Creative Commons licenses have been "ported" over 50 different jurisdictions worldwide. No new ports are being started as preparations for version 4.0 of the license suite begin.[1]

Original licenses

The original set of licenses all grant the "baseline rights", such as the right to distribute the copyrighted work worldwide, without changes, at no charge.[2] The details of each of these licenses depends on the version, and comprises a selection of four conditions:

Attribution (by)	Licensees may copy, distribute, display and perform the work and make derivative works based on it only if they give the author or licensor the credits in the manner specified by these.
Noncommercial (nc)	Licensees may copy, distribute, display, and perform the work and make derivative works based on it only for noncommercial purposes.
No Derivative Works (nd)	Licensees may copy, distribute, display and perform only verbatim copies of the work, not derivative works based on it.
Share-alike (sa)	Licensees may distribute derivative works only under a license identical to the license that governs the original work. (See also copyleft.)

Combinations

Mixing and matching these conditions produces sixteen possible combinations, of which eleven are valid Creative Commons licenses and five are not. Of the five invalid combinations, four include both the "nd" and "sa" clauses, which are mutually exclusive; and one includes none of the clauses. Of the eleven valid combinations, the five that lack the "by" clause have been retired because 98% of licensors requested attribution, though they do remain available for reference on the website.[3] [4] [5] This leaves six regularly used licenses:

- Attribution alone (by)
- Attribution + NoDerivatives (by-nd)
- Attribution + ShareAlike (by-sa)
- Attribution + Noncommercial (by-nc)
- Attribution + Noncommercial + NoDerivatives (by-nc-nd)
- Attribution + Noncommercial + ShareAlike (by-nc-sa)

For example, the Creative Commons Attribution (BY) license allows one to share and remix (create derivative works), even for commercial use, so long as attribution is given.[6]

Attribution

Since 2004, all current licenses require attribution of the original author.[4] The attribution must be given to "the best of [one's] ability using the information available".[7] Generally this implies the following:

- **Include any copyright notices (if applicable)**. If the work itself contains any copyright notices placed there by the copyright holder, those notices must be left intact, or reproduced in a way that is reasonable to the medium in which the work is being re-published.
- **Cite the author's name, screen name, or user ID**, etc. If the work is being published on the Internet, it is nice to link that name to the person's profile page, if such a page exists.
- **Cite the work's title or name (if applicable)**, if such a thing exists. If the work is being published on the Internet, it is nice to link the name or title directly to the original work.
- **Cite the specific CC license the work is under**. If the work is being published on the Internet, it is nice if the license citation links to the license on the CC website.
- **Mention if the work is a derivative work or adaptation**, in addition to the above, one needs to identify that their work is a derivative work i.e., "This is a Finnish translation of [original work] by [author]." or "Screenplay based on [original work] by [author]."

Applicable works

Work licensed under a Creative Commons License is governed by applicable copyright law.[8] This allows Creative Commons licenses to be applied to all work falling under copyright, including: books, plays, movies, music, articles, photographs, blogs, and websites. Creative Commons does not recommend the use of Creative Commons licenses for software.[9]

However, application of a Creative Commons license may not modify the rights allowed by fair use or fair dealing or exert restrictions which violate copyright exceptions. Furthermore, Creative Commons Licenses are non-exclusive and non-revocable. Any work or copies of the work obtained under a Creative Commons license may continue to be used under that license.

In the case of works protected by multiple Creative Common Licenses, the user may choose either.

Other licenses

A number of additional licenses have been introduced, which are more specialized:

- Sampling licenses, with two options:
  - **Sampling Plus**: parts of the work can be copied and modified for any purpose other than advertising, and the entire work can be copied for noncommercial purposes[10]
  - **Noncommercial Sampling Plus**: the whole work or parts of the work can be copied and modified for noncommercial purposes[11]

Retired licenses

Due to either disuse or criticism, a number of previously offered Creative Commons licenses have since been retired,[12] and are no longer recommended for new works. The retired licenses include all licenses lacking the Attribution element[3] other than CC0, as well as two licenses not allowing at least non-commercial verbatim distribution worldwide:

- **Sampling**: parts of the work can be used for any purpose other than advertising, but the whole work cannot be copied or modified
- **DevNations**: a Developing Nations license, which only applies to countries deemed by the World Bank as a "non-high-income economy". Full copyright restrictions apply to people in other countries.
- **Sampling Plus**

Public domain tools

Besides licenses, Creative Commons also offers a way to release material into the public domain through CC0,[13] a legal tool for waiving as many rights as legally possible, worldwide. Development of CC0 began in 2007[14] and the tool was released in 2009.[15] [16]

In 2010, Creative Commons announced its Public Domain Mark,[17] a tool for labeling works already in the public domain. Together, CC0 and the Public Domain Mark replace the *Public Domain Dedication and Certification,* [18] which took a U.S.-centric approach and co-mingled distinct operations.

In 2011, Free Software Foundation added CC0 to its free software licenses, making CC0 a recommended way of dedicating software to the public domain.[19]

Partial list of projects that release contents under Creative Commons licenses

- Association for Progressive Communications (cc-by-sa)
- ccMixter (mostly cc-by-nc)
- Citizendium (cc-by-sa)
- The Freesound Project (Sampling Plus)
- Identi.ca (cc-by)
- knol (mostly, cc-by-sa or cc-by-nc-sa)
- Mushroom Observer (cc-by-sa or cc-by-nc-sa)
- Open Game Art (CC-By and -SA 3.0 without NC, CC0)
- Wikia (cc-by-sa, since June 2009)
- Wikinews (cc-by)
- Wikipedia (cc-by-sa, since June 2009)
- Wikitravel (cc-by-sa)

See also

- Creative Commons
- Creative Commons jurisdiction ports
- Copyright
- Non-commercial educational
- Free Software
- Free Music
- Free Culture

References

[1] "CC Affiliate Network" (http://wiki.creativecommons.org/CC_Affiliate_Network#The_Licensing_Suite). Creative Commons. . Retrieved July 8, 2011.
[2] "Baseline Rights" (http://wiki.creativecommons.org/Baseline_Rights). Creative Commons. June 12, 2008. . Retrieved February 22, 2010.
[3] "Retired Licenses" (http://creativecommons.org/retiredlicenses). Creative Commons. . Retrieved July 5, 2007.
[4] Announcing (and explaining) our new 2.0 licenses (http://creativecommons.org/weblog/entry/4216)
[5] "Creative Commons Licenses" (http://creativecommons.org/licenses/). Creative Commons. . Retrieved February 22, 2010.
[6] "Creative Commons — Attribution 3.0 United States" (http://creativecommons.org/licenses/by/3.0/us/). Creative Commons. November 16, 2009. . Retrieved February 22, 2010.
[7] "Frequently Frequently Asked Questions" (http://wiki.creativecommons.org/FFAQ#How_do_I_properly_attribute_a_Creative_Commons_licensed_work.3F). Creative Commons. February 2, 2010. . Retrieved February 22, 2010.
[8] "Creative Commons Legal Code" (http://creativecommons.org/licenses/by/2.5/au/legalcode). Creative Commons. January 9, 2008. . Retrieved February 22, 2010.
[9] Creative Commons FAQ: Can I use a Creative Commons license for software? (http://wiki.creativecommons.org/FAQ#Can_I_use_a_Creative_Commons_license_for_software.3F)
[10] "Sampling Plus 1.0" (http://creativecommons.org/licenses/sampling+/1.0/). Creative Commons. November 13, 2009. . Retrieved February 22, 2010.
[11] "Creative Commons — NonCommercial Sampling Plus 1.0" (http://creativecommons.org/licenses/nc-sampling+/1.0/). Creative Commons. November 13, 2009. . Retrieved February 22, 2010.
[12] Lessig, Lawrence (June 4, 2007). "Retiring standalone DevNations and one Sampling license" (http://creativecommons.org/weblog/entry/7520). Creative Commons. . Retrieved July 5, 2007.
[13] "CC0" (http://creativecommons.org/choose/zero). Creative Commons. . Retrieved February 22, 2010.
[14] "Creative Commons Launches CC0 and CC+ Programs" (http://creativecommons.org/press-releases/entry/7919) (Press release). Creative Commons. December 17, 2007. . Retrieved February 22, 2010.
[15] Baker, Gavin (January 16, 2009). "Report from CC board meeting" (http://www.earlham.edu/~peters/fos/2009/01/report-from-cc-board-meeting.html). *Open Access News*. . Retrieved February 22, 2010.
[16] Expanding the Public Domain: Part Zero (http://creativecommons.org/weblog/entry/13304)
[17] Marking and Tagging the Public Domain: An Invitation to Comment (http://creativecommons.org/weblog/entry/22940)
[18] "Copyright-Only Dedication (based on United States law) or Public Domain Certification" (http://creativecommons.org/licenses/publicdomain/). Creative Commons. August 20. 2009. . Retrieved February 22, 2010.
[19] "Using CC0 for public domain software" (http://creativecommons.org/weblog/entry/27081). Creative Commons. April 15. 2011. . Retrieved May 10, 2011.

External links

- Official site (http://www.creativecommons.org/)
- Full selection of licenses (http://www.creativecommons.org/about/licenses/)
- Creative Commons Licenses Compatibility Wizard (http://www.creativecommons.org.tw/licwiz/english.html)
- Citizen's guide to using Creative Commons licenses (http://www.newmediarights.org/guide/legal/copyright/licensing/creative_commons/citizens_legal_guide_using_creative_commons_licenses)
- A human readable summary of version 2.5 of the Creative Commons CC-BY-SA license. (http://freedomdefined.org/Licenses/CC-BY-SA-2.5) (Retrieved 2010-04-10).
- Legal code of the CC-BY-SA 3.0 License. (http://creativecommons.org/licenses/by-sa/3.0/legalcode)

Blimp

The Spirit of Goodyear, one of the iconic Goodyear Blimps

A **blimp**, or **non-rigid airship**, is a floating airship without an internal supporting framework, or keel. A non-rigid airship differs from a semi-rigid airship and a rigid airship (e.g., a Zeppelin) in that it does not have any rigid structure, neither a complete framework nor a partial keel, to help the airbag maintain its shape. Rather, these aircraft rely on both a higher pressure of the lifting gas (usually helium) inside the envelope and the strength of the envelope itself.

The term "blimp" refers only to free-flying aircraft. The term is sometimes erroneously used to refer to the tethered craft known as moored balloons. While often very similar in shape, moored balloons have no propulsion and are tethered to the ground.

Principle

Steerable ducted fans on a Skyship 600 provide thrust, limited direction control, and also serve to inflate the ballonets to maintain the necessary overpressure

Since blimps keep their shape with internal overpressure, typically the only solid parts are the passenger car (gondola) and the tail fins. A non-rigid airship that uses heated air instead of a light gas (such as helium) as a lifting medium is called a hot-air airship.

Volume changes of the lifting gas, due to temperature changes, is balanced using ballonets (air bags), in order to maintain the overpressure. Without sufficient overpressure, the blimp loses its ability to be steered and top speed is degraded. The propeller air stream can be used to inflate the hull. In some models, such as the Skyship 600, differential ballonet inflation can provide a measure of pitch trim control.

The engines driving the propellers are usually directly attached to the gondola, and in some models are partly steerable.

Blimps are the most commonly built airships, because they are relatively easy to build and easy to transport once deflated. However, because of their unstable hull, their size is limited. A blimp with too long a hull may kink in the middle when the overpressure is insufficient or when maneuvered too fast (this has also happened with semi-rigid airships with weak keels). This led to the development of semi-rigids and rigid airships.

Modern blimps launch somewhat heavier than air (overweight), in contrast to historic blimps. The missing lift is provided by lifting the nose and using engine power. Some types also use steerable propellers or ducted fans. Operating in a state heavier than air avoids the need to dump ballast at lift-off and also avoids the need to lose costly lifting gas on landing.

Etymology

The term "blimp" is reportedly onomatopoeic, the sound the airship makes when one taps the envelope (balloon) with a finger.[1] Although there is some disagreement among historians, credit for coining the term is usually given to Lt. A.D. Conningham of the British Royal Navy in 1915.[2]

A modern blimp from Airship Management Services showing a strengthened nose, ducted fans attached to the gondola under the hull, and cable-braced fins at the tail

A 1943 etymology published in the *New York Times* confirms the British origin during the First World War when the British were experimenting with lighter-than-air craft. The initial non-rigid aircraft was called the A-limp; and a second version called the B-limp was deemed more satisfactory.[3]

A different derivation is given by Barnes & James in *Shorts Aircraft since 1900*:

> In February 1915 the need for anti-submarine patrol airships became urgent, and the Submarine Scout type was quickly improvised by hanging an obsolete B.E.2c fuselage from a spare Willows envelope; this was done by the R.N.A.S. at Kingsnorth, and on seeing the result for the first time, Horace Short, already noted for his very apt and original vocabulary, named it "Blimp", adding, "What else would you call it?"[4]

An oft-repeated, but false, etymology is that the United States military initially had two classes of aircraft: A-rigid and B-limp, hence "blimp". In fact,

> ... there was no American "A-class" of airships as such – all military aircraft, heavier or lighter-than-air were designated with "A" until the appearance of B-class airships in May 1917. There was an American B airship – but there seems to be no record of any official designation of non-rigids as "limp". Further, according to the Oxford Dictionary, the first appearance of the word in print was in 1916, in England, a year before the first B-class airship.
>
> —"Etymology of "Blimp" by Dr. A. D. Topping, *AAHS Journal*, Winter 1963.

Use

The B class blimps were patrol airships operated by the United States Navy during and shortly after World War I. The Navy learned a great deal from the DN-1 fiasco. The result was the very successful B-type airships. Dr. Jerome Hunsaker was asked to develop a theory of airship design, Lt. John H. Towers returned from Europe having inspected British designs, the Navy sought bids for 16 blimps from American manufacturers. On 4 February 1917 the Secretary of the Navy directed that 16 nonrigid airships of Class B be procured. [1] Ultimately Goodyear built 9 envelopes, Goodrich built 5 and Curtiss built the gondolas for all of those 14 ships. Connecticut Aircraft contracted with U.S. Rubber for its two envelopes and with Pigeon Fraser for its gondolas. The Curtiss-built gondolas were modified JN-4 fuselages and were powered by OX-5 engines. The Connecticut Aircraft blimps were powered by Hall-Scott engines.

In 1930 a former German airship officer, Captain Anton Heinen, working in the US for the US Navy on it dirigible fleet attempted to design a build a four place blimp called the **family air yacht** for private fliers which if placed in production the inventor claimed would be priced below $10,000 and easier to fly than a fixed wing aircraft. It was unsuccessful. [5] [6]

Examples of non-rigid airships

There are several blimps worldwide.[7] Some examples include:

- TC-3 and Tc-7, two US Army Corps non-rigid blimps used for parasite fighter trials during 1923–24.
- SS, SSP, SST, SSZ and NS class blimps, convoy escort blimps used by the UK in World War I.
- G class blimp and L class blimp, US training blimps built by Goodyear during World War II.
- K class blimp and M class blimp, US anti-submarine blimps operated during World War II.
- N class blimp (the "Nan ship"), used for anti-submarine and as a radar early-warning platform during the 1950s.
- Goodyear Blimps, a fleet of blimps operated for advertising purposes and as a television camera platform.
- Skyship 600, a private blimp used by advertising companies
- P-791, an experimental aerostatic/aerodynamic hybrid airship developed by Lockheed-Martin corporation.
- SVAM CA-80, an airship manufactured by the Shanghai Vantage Airship Manufacture Co in China

DIRECTV Airship, Model A-170LS Video Lightsign by American Blimp Corporation

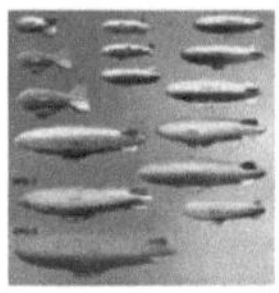

An American Blimp Corporation A-60+, the MetLife Snoopy Two[8]

See also

- Airship hangar
- Hindenburg class airship
- Mooring mast

Notes

[1] Myers, Steven Lee. "A Busted Blimp Releases a Large Giggle Factor," (http://www.nytimes.com/1993/07/11/weekinreview/a-busted-blimp-releases-a-large-giggle-factor.html?scp=1&sq=A Busted Blimp Releases a Large Giggle Factor&st=cse) *New York Times.* July 11, 1993.

[2] For example, in:

- Dickson, Paul (2004). *War Slang: American Fighting Words and Phrases Since the Civil War.* Brassey's. pp. 42 (http://books.google.com/books?id=c-Xv836lmxgC&pg=PA42)– 43 (http://books.google.com/books?id=c-Xv836lmxgC&pg=PA43). ISBN 978-1-57488-710-5.
- *The Aeronautical Journal* (Royal Aeronautical Society) **72**: p. 857. 1968. ISSN 0001-9240. (See snippets 1 (http://books.google.com/books?id=i8zFAAAAIAAJ&q=blimp+cunningham) and 2 (http://books.google.com/books?id=i8zFAAAAIAAJ&q=blimp) on Google Books.)

[3] "Origin of 'Blimp' Explained," *New York Times.* January 3, 1943

[4] Barnes & James, p.13

[5] "Aeronautics: Air Yacht" *Time Magazine*, November 3rd 1930 (http://www.time.com/time/magazine/article/0,9171,882391,00.html)

[6] "Dirigible Air Yacht Has Automobile Cabin" *Popular Mechanics*, December 1930 (http://books.google.com/books?id=qOIDAAAAMBAJ&pg=PA967&dq=Popular+Science+1930+plane+"Popular+Mechanics"&hl=en&ei=_7BITsWeBYTWgQf9mIiLCg&sa=X&oi=book_result&ct=result&resnum=10&sqi=2&ved=0CE8Q6AEwCQ#v=onepage&q&f=true)

[7] "FAQs – Business of blimps" (http://www.goodyearblimp.com/faqs/faqs_business.html#three). Goodyear Tire and Rubber Company. . Retrieved 2009-12-13.

[8] "The MetLife Blimp" (http://www.metlife.com/about/advertising/metLife-blimp/index.html). MetLife. 2009. . Retrieved 2009-12-13.

References

- Barnes C.H. & James D.N. *Shorts Aircraft since 1900.* London (1989): Putnam. pp. 560. ISBN 0-85177-819-4.

External links

- *Popular Mechanics*, June 1943, *"Gas Bags Go On Patrol"* (http://books.google.com/books?id=bNYDAAAAMBAJ&pg=PA34&dq=popular+mechanics+1943+flying+shark&hl=en&ei=XFeQTOPoGorCnAeW18i0DA&sa=X&oi=book_result&ct=result&resnum=1&ved=0CCsQ6AEwAA#v=onepage&q=popular mechanics 1943 flying shark&f=true) detailed article on antisubmarine blimps during World War Two
- "How The First Sea-Air Rescue Was Made", October 1944, Popular Science (http://books.google.com/books?id=PyEDAAAAMBAJ&pg=PA114&dq=motor+gun+boat&hl=en&ei=LxTqTMfGI4-bnwfEyNiWDQ&sa=X&oi=book_result&ct=result&resnum=6&ved=0CEIQ6AEwBQ#v=onepage&q=motor gun boat&f=true) first air-to-sea rescue without aircraft landing first

Article Sources and Contributors

NearMap *Source*: http://en.wikipedia.org/w/index.php?title=NearMap *Contributors*: Colin Marquardt, Firefishy, Grutness, Kocio, Lakeyboy, MoondyneAWB, Rich Farmbrough, Swoof, TRS-80, Visik, 7 anonymous edits

2010–2011_Queensland_floods *Source*: http://en.wikipedia.org/w/index.php?title=2010%E2%80%932011_Queensland_floods *Contributors*: Active Banana, Adamdaley, Adpete, Aeonx, Agiseb, Aiden Artyo More, Aldo samulo, Anonymous Dissident, Arla, Ashmoo, AstroHurricane001, Australia2world, Awa655, Axolotl Nr.733, Azumanga1, Beao, BenJWoodcroft, Bencherlite, Bender235, Benforward85, Bfryar, Biatch, Bidgee, Bilby, Billben12, Bitstomper, Bluefist, Bowlderizer, Brisbane Auth, Bueller 007, Cablehorn, Calle Widmann, CanberraBulldog, Chansonh, Christian75, Csmart287, Cybrspunk, Cyclonebiskit, DMacks, DVdm, Danallen46, Danbarnesdavies, DermottBanana, Diannaa, Diego Grez, Djackmanson, Dmol, Don4of4, Durex24, El T, Enidblyton11, Eric Kvaalen, Ericoides, Falcor84, Fanatix, Fnielsen, Forteller, Gogo Dodo, Grafen, Ground Zero, Guoguo12, HJ Mitchell, HalloranElder, Hbdragon88, Hewhoamareismyself, HiLo48, Hinata, Honeymanl, Hurricanehink, Ianblair23, Jack Merridew, JackofOz, Jameshbailie, Jamfrank, Jason Rees, JediJeremy, Jimbo online, Jimp, Jjron, John Vandenberg, Johnsemlak, Johnson487682, JustAGal, Kenatipo, Kevinmon, Khazar, Khitty, Kilo-Lima, Kno0006, Knowz, Koavf, Kwamikagami, Kylelovesyou, Lankiveil, Leigh, LilHelpa, LowKey, Lugnuts, Luke Thorly, M.O.X, MJCdetroit, Many baks, Marek69, Materialscientist, Matthiasb, Mattinbgn, McAusten, Mclaren2008, MelbourneStar, Melburnian, Mervyn, Mfpvehicles, Miblo, Miesianiacal, Mike Rosoft, Moondyne, MoondyneAWB, Mymy101, N01 hobo, NJM2010, Nhyty, Nick-D, Nmk829, Nohomers48, OfficeBoy, Ohconfucius, OlEnglish, Optimist on the run, Orange Suede Sofa, Orderinchaos 2, Originalwana, Ozkidzez91, PMBO, PRRfan, Peaceworld111, Pickles8, Portillo, Ppntori, Prnguin, R'n'B, Rangasyd, Reaper Eternal, Rocketrod1960, Rohzan, Roidroid, Rossrs, Ryan White Jr., Saffron09, SamB135, SatuSuro, Servimus, Shaun2212, Shiftchange, ShipFan, Smegs07, Smithrooks, Soap, Someoneawesomlycool, Sp33dyphil, Sparkzy, StAnselm, Suid-Afrikaanse, Tbhotch, Teagan.kinging, Thatsgold, The-Pope, Thedeanjones, Themeparkgc, Tide rolls, Titoxd, Tmangray, Tri400, Triamks, Truthanado, Tymmee, Uaumer, Vegas949, Vgy7ujm, Vinithehat, WLRoss, WWGB, Waddles3900, Waffles4509, Ya hui chou, Yellowdesk, Your Lord and Master, Your Next Kid, Романов, 309 anonymous edits

Ipswich,_Queensland *Source*: http://en.wikipedia.org/w/index.php?title=Ipswich%2C_Queensland *Contributors*: -1g, 32X, 5fingers, Abcdefghayden, Ah48, Alex Law, Amj45, Anotherclown, Ansell, Aranel, BRUTE, Barrypaton, Bastique, Berichard, Bestywins, Bgwhite, Biatch, Bidgee, Bigisbig, Bogan123, Bongomanrae, Booba1994, BrightLights, Brizboy90, CEP78, CJ, Calaiskid, Camerong, Can't sleep, clown will eat me, CanadianLinuxUser, CapitalR, Cgoodwin, Chanvw, Ched Davis, Chickenchippies, Choij, Chris j wood, Chris86hull, Clarkk, Cmdrjameson, Colin MacLaurin, Craftyminion, Crusoe8181, Ctatt6, Danmiles, Danny, DorianGray, Drag.15uzu, Dysprosia, EmptyV, Eregli bob, Eugene-elgato, Everyking, Figaro, Firsfron, Fuzheado, GRider, Gene Nygaard, Geoking66, Gmdillon, GoingBatty, Goodstone, Graham87, Grahamec, Ground Zero, Gurch, HPootter, HangingCurve, Harryboyles, Hephaestos, Hersfold, Hmains, Humehwy, ImperialAssassin, Itsmeagain101, JV97, James086, Jevansen, John Vandenberg, Johnmc, Joseph Solis in Australia, JyeBurton, Kaleb777, Katieh5584, Kayla15, KazukiMiyajima, Kotniski, Kyruph, Lankiveil, Lav90, Lepardion, Lindblum, Liveste, LordPhantomAU, Lovemyrice, LowKey, MagpieShooter, Mathieumcguire, Mattinbgn, Mboverload, Mcoupal, Melody, Michaelbusch, Michellecrisp, Mike Rosoft, Mindmatrix, Miracle Pen, Mlaffs, Mpidg3, Mvjs, NPrice, NewAust, Nick Number, Nick-D, Nika 243, Niteowlneils, Nosferatu.feuertote, Open2universe, Orderinchaos, Orderinchaos 2, Orodreth, Oxymoron83, Ozurbanmusic, PDH, Paul A, Peripitus, PhilKnight, Philc0005, Pirate Clarke, Plastikspork, Reaper Eternal, RepublicanJacobite, Retired username, Rich Farmbrough, Rob Lindsey, Robert Brockway, Roleplayer, Ross Grove, Rulesfan, SM247, Sb617, ScottDavis, Serp3821, Shiftchange, Simply south, Skysmith, Smallestg, Smickel, Soozlepip, Spewmaster, Springfieldboy, Sting au, Tallegalla, Tamara1503, Tanglewood4, Tee4306, Theshadow1990, Triathalon, Triddle, Troycook, Ulric1313, Vannasouk, Wannabcommedian, Wouterhagens, YellowMonkey, 282 anonymous edits

OpenStreetMap *Source*: http://en.wikipedia.org/w/index.php?title=OpenStreetMap *Contributors*: Actiuinformatica, Advanstra, Alexrudd, Altermike, And ypsilon, Antimirov, Apmon, Aranymalinko, Artemgy, Atallcostsky, Avenue, Axrtest, Bahar101, Balrog-kun, Ben robbins, Bender235, Benlast, Bevo, Binjiangwiki, Bruce89, Brupp, Budyakov, Bumbulski, Bus88MRT, Canley, Ceyockey, ChaRleyTroniC, CharlesC, Chendy, ChristianVetter, Ckatz, Coderminus, Combbs, Compdude123, Cybiko123, D6, Dandv, Danim, Dankarran, Davespod1, Derek farn, Doubleyouyou, Dpm64, Dream out loud, Dwi Secundus, Edward, Efa, Egil, Elgaard, EugeneZelenko, Evgen Bodunov, Ewald, Firefishy, Fleminra, Flip, Fmph, Frap, Gobonobo, GrahamAsher, Grand Edgemaster, Grendelkhan, Gronky, Grotte, Guyjohnston, Hajimegane, Harry Wood, Hawky, Hobsonlane, Horkana, Hpt, Hsnavigator, Hugo75, Hydrargyrum, Ian Moody, Ida Shaw, Imc, IrrtNie, Iulian.serbanoiu, Iulianu, Iusethis, Ivansanchez, JLaTondre, Jaan513, Jameslwoodward, Jammeh, Jeff G., Jidanni, Jldominguez, Jmartama, Jonobennett, Jrouquie, Justjkk, Jw4nvc, Jynus, KAMiKAZOW, Karnesky, Kathleen.wright5, Kenguest, Kocio, Koying, Kozuch, Kuru, LA2, LMB, Lquilter, Lucid, Luen, Lurker, MCWNT, Marcus Qwertyus, Markiliffe, Martarius, Maurreen, Maxim, Mcld, Mike Linksvayer, Mindmatrix, Mintywalker, Mms, Mond, Mortenoesterlundjoergensen, Mortense, Mosca, Mr.Z-man, Muro de Aguas, Mwtoews, NeMeSiS94, Nfgusedautoparts, NicD, Nickblack, Nigelj, Nihadkav, Nikai, NittyG, Nkosi, Noformation, Nroets, Oijkoijk, Ojw, Oliverlewis, Ottre, Padraic, Pavol.zibrita, Pegua, Peter E. James, PeterEastern, PeterIto, Petr Dlouhý, Pigsonthewing, Ptigrouick, Quatloo, QuintS, R'n'B, Random832, Raymond, Rhkramer, Richardprins, Rjwilmsi, Rmac8, Roland Hansson, Rovastar, Rural Bengali, Rurseekatze, RussNelson, Samanthaenvgov, Shane28r, Sissssou, Sk5893, Skizzik, Sladen, Sleske, Smsm1, Snorre, Socrates2008, Softeis, Southglos, Speedevil, StefanB, Stephenchou0722, Stuuf, T0ny, TarzanASG, Techtoucian, Teratornis, Thrapper, Thumperward, Tlogmer, Tmcw, Tobias K., TomArra, Tony Rotondas, Ulric1313, Unxed, UrusHyby, Vclaw, W-j-s, WJetChao, Waldir, Welsh, William Avery, Wizardist, Woohookitty, WookieInHeat, Wyrjk, X-mass, Xan2, Xiong, Ynhockey, ZeroOne, Zorankovacevic, Милан Јелисавчић, 220 anonymous edits

Aerial_photography *Source*: http://en.wikipedia.org/w/index.php?title=Aerial_photography *Contributors*: (;D- hymyilevä takapiru, A. B., APh, Aerialphotopros, Aerialscaper, Allanrobertwilliams, AndrewDressel, Anlace, Arpingstone, Avalerion, Avibrazil, AxelBoldt, Beechhouse, Benea, Bietje, Blimpp, BlueCanoe, Bluemoose, Bodybagger, Buffs, Camillaw, Cdinesh, Ceyockey, Charlesdrakew, Chase me ladies, I'm the Cavalry, Chmod007, ChristopherBorcsok, Coll7, CvetanPetrov1940, DJ Clayworth, Danhash, Danielchameleon, Davewarwick, DavidFarmbrough, Dcaps, Ddxc, Deathoffilmnoir, Deli nk, Dhaluza, Dicklyon, Dpakdel, Duja, Eastlaw, Edsanville, Eve Hall, Ewingmh, Extra999, Eyeintheskye, FGLawson, Fhbwgads, Filtv, Flamelai, Fleckb8531, Flockmeal, Fotonix photo, Foxandpotatoes, Gabriel Kielland, Gadfium, George The Dragon, Georgejdorner, Glenn, Globalflyer, Goodsandsrvcs, Gunter, Hankwang, Hans Adler, Higherview, Hogne, Hooperbloob, Hotlorp, Hu, Hugo999, ILFoxtrot, Iszdfb, Ivansanchez, JRoberts72, Jackehammond, Jmajoris, JoeB52, John Comeau, JuanFox, Justicelovespeace, Jw4nvc, KJG2007, Kbosak, Kitegibsongirl, Kprateek88, LeaW, Legifotok.hu, Lightmouse, Luna Santin, MSGJ, Mactographer, MadameSarah, Maggiek2, Markus Kuhn, MarylandArtLover, Matheushm, Maurice J, Mblx6jr2, Memphis101, Michael Shields, Mike.lifeguard, My name is al, Mywood, Natcase, Nihil novi, Nposs, Nuty350, Ohconfucius, Ohnoitsjamie, Ozzmosis, PFHLai, PMDrive1061, Patrick, Personalaccount, Phantom 71, Phatmonkey, Phgao, Picaroon, Pixaerial, Postdlf, R'n'B, Rcc105, Recury, Redjar, Redthoreau, Retired username, Rjm at sleepers, Rlandmann, Rmckay630, RobC450, Rohan89, Roland zh, Ronk01, Russavia, SGGH, Sam Korn, Sapientum, Satori Son, Scriberius, Scwlong, Secretlondon, Seotommy, ShutterBugTrekker, Sprocket, Steel, Stevieggg01, Stonor, StuffOfInterest, Superzoulou, Susshma, Syrthiss, Tacoje7, Taosupremo, TastyPoutine, TheEgyptian, TheStripèdOne, Tntdj, Tsamuels1, Van helsing, Vegaswikian, Veinor, Velella, Wangi, Webquest, WikiDao, Woohookitty, Wtshymanski, Ww2censor, Y23, Zigger, Zonk43, Zscout370, يامان, 187 anonymous edits

Digital_elevation_model *Source*: http://en.wikipedia.org/w/index.php?title=Digital_elevation_model *Contributors*: Aarchiba, Abdull, Alanfeynman, Aleichem, AlexD, Altiusfortius, Altmany, Alvaro, Andreygeo, Armysurveyor, Bdiscoe, Bobblewik, CDN99, Checki J Fischer, Chris the speller, Clemifornia, Damian Yerrick, DaveDixon, Eve Hall, Eyeintheskye, Fhelmli, Frodet, Gahs, Geologicharka, Gulliveig, Hans Braxmeier, Hengl, Jeffaudi, Jwo, Kbh3rd, Kbosak, LjL, Lvhengani, Manop, MartinOver, Maurice J, Mav, Mejor Los Indios, Mwtoews, MyNameIsNeo, Nagualdesign, Natcase, O'Dea, Plasma east, PlatypeanArchcow, Quantumobserver, R'n'B, RP459, Radmap, Raven in Orbit, Razor2988, Redjar, Rich Farmbrough, Robertwhitinger, Ronz, STLocutus, Simonchalky, SoilMan2007, Strebe, THEN WHO WAS PHONE?, Tddecker, Ultrasonicspeed, Viewfinder, Whitejay251, Wikianon, Winterst, World of Radar, Yubal, Zhangmoon618, 65 anonymous edits

Brisbane *Source*: http://en.wikipedia.org/w/index.php?title=Brisbane *Contributors*: $eti, -- April, 1ctv818, 203.0.230.xxx, 210.215.40.xxx, AQjosh, Aaron Stitz, Aaroncrick, Addshore, Adz, Aggelophoros, Aggressivesecularist, Ahoerstemeier, Aioth, Akrabbim, Alai, Alan Liefting, Alandavidson, AlbertusIII, Alex Law, Alexbrabyn, Alexrg2003, Aliasd, All stroked out on julian, Altormainstream, Amakuru, Amun1, Anaxial, Andrea105, AndrewHowse, Andrewtechhelp, Andy Marchbanks, Andyabides, Andyman14, Angel2001, Angela26, Anoldtreeok, Anotherclown, Ansell, AppleJuggler, Arashi, ArglebargleIV, Arjun01, Ascough, Ashmoo, Ashton 29, Asphinx, Astro1, AtD, Atlantis Hawk, Auriong, Aus Roads, AussieLegend, AussiePS, Australia boy, Avala, Avnjay, Awa655, AxSkov, AxelBoldt, Azeari, Azumanga1, BEZALEL2000, Babygoanna, Bahadorjn, Banserki, BarbD, Barek, Bastian92, BazookaJoe, Belfunk, Beneaththelandslide, BennyD11, Berlinerzeitung, Best name, Biatch, Bidgee, BigHaz, Bilious, Billyb, Bjdorge, BjornVDM, Bkdd, Blackknight12, BlaiseFEgan, Blanchardb, Bleakcomb, Bloodpack, Bob Palin, Bobblewik, Bobo192, Bongomanrae, Bongwarrior, BooblIaAU, BrandonsLe, Breezer84, Bretto, Briancua, Brianj20, BrightLights, Brighterorange, Brisbane Man, Brisbane7, BrisbanePom, Brisbanite, Brisvegas, Briz18, Brizboy90, Brizbun, Broncos9900, Bronwynforster, Brískelly, Bucoli, Budley B, Bufan123, Burgundavia, CIreland, CJ, CWY2190, Caiaffa, Caknuck, CambridgeBayWeather, Canley, Canuck85, CapnZapp, CaptainJSparrow, Carnun, CarpetCello, Casliber, Cassowary, Catgut, Celcom, Cenarium, Cerberus584, Challenge me, Chameleon, Chanvw, Chimpa, ChiniMick2005, Chinimick, Chlorem, Chris the speller, Chris.Giles, ChrisW, Chriskong37, Churba, Ck786, Ck79247, Ckatz, Clancycol, Clare94, Clarkk, Cmdrjameson, Cnbrb, Cntras, Coffee, Colin MacLaurin, Cometstyles, Commander Keane, CommonsDelinker, Conscious, Conversion script, Cool blade, Craig Sunderland, CrazyC83, CrazyChemGuy, Crusoe8181, Cs-wolves, Curieux, Curps, Cvene64, Cyndi88, D'oh!, D-Notice, D6, DARTH SIDIOUS 2, DC text, DChang88, DVdm, DabMachine, Dabomb87, Dalaskan, Dallasfan11, Dalziel 86, Danc, Daniel Callegaro, Daniel Yasar, Danr2k6, Danubis, Dark Lord of the Sith, Darrenfong28, Darwen, Daven200520, Dburr, Dcollyer, Deathcabforbunny, Dechetes, Dedark, Delirium, Delldot, Denkealsobin, Depor23, Deror avi, Deselliers, Diceman, Diglibs2, Diliff, Dimethyl, Dina, Djreptar, Dmol, DocendoDiscimus, Dpbourke, Dr Ailsa, Dr Marty, Drawley, Drjt87, Dryazan, Duncanferg, Dushan wideway, Dvavasour, Dwstein, Easel3, Effer, El C, Elekhh, Elmonoro, Emathias, Ember of Light, EmersonLowry, Empirevision, Eno1, Entropy1963, Eric Shalov, Erich gasboy, Erik9, Evans1982, Evanstim, Evercat, Exemplar sententia, Ezykron, EzykronHD, FNQ, Factoid Killer, Falcadore, Faradayplank, Faustbol, Figaro, FiggyBee, Finlay McWalter, Finneganw, Flewis, Floria L, Floul1, Forever Queensland, Fox, Franamax, Frasharv123456789, Frogskin9, Fudoreaper, G.hartig, GB fan, GSlicer, Gadfium, Gaius Cornelius, Galati, Gatkinson, Gaudoine, Gavin Moodie, Gaz, Gcampton, Genconian, Gene Nygaard, Georg Amadé, Ghost 9, Giggy, Gigpig, Gilliam, Glen 1197, Glibik, Gnangarra, GoQueensland, Goatsmilk, Gobeirne, Goldom, Gota 93, Grace Note, Graham87, Grant65, Grstain, Gurch, H3802.1266, HERB, HPL1432, Hadal, Hamedog, Hamwheel1, HappyWaldo, Harbison, Harryboyles, HashBrowns&Beans, Hayabusa future, Hazbaz, Hephaestos, Hertz1888, HiLo48, HitroMilanese, Holasenorunitedstates, Homelessblueless, Honette, Huge Bananas, Humehwy, Hushpuckena, Hustler001, Ianblair23, Ikiroid, Infrogmation, Interiot, Iridescent, Ironchef90, IsabelleF, Iscah, IslandGyrl, J Herington, J Milburn, J.delanoy, JLaTondre, JPD, JT72, JaGa, Jacinta2003, Jackorino, Jackp, Jacksav, Jaems, Jajr, James Pinnell, JamesR, Jauhienij, Jay Kak, JayKeaton, Jeff79, Jeffrey shaw, Jeffro77, Jemijohn, Jeronim, JerryHstevensjr, Jgmdownloads, Jiang, Jimboclone, Jimfbleak, Jimp, Jivecat,

Jjdevoy, Jocelynaus, Jockosaurus, John Vandenberg, John of Reading, John1abbes, JohnCD, Johnl, JohnOwens, Johnmc, Johnnye21 21 21, Jonathan.s.kt, Jonik, Jose77, Joseph Solis in Australia, Joshd, Joshii, Joshisawesome1000, Jpbowen, Jpp42, Just James, Jv821, Jvblackley, Jyril, KFP, Kai11, Kakofonous, Kalmia, Karen Johnson, Karyasuman, Kate, Kate.woodcroft, Katerina234234, Katty121, Kazfernandes, Kazzaw, Keladino, Kerry Raymond, Kgbo, Kikvadze03, Kirkbroadhurst, Klippa, Kmcnamee, Koorja, Koyaanis Qatsi, Kransky, Kungfuadam, Kurowoofwoof111, Kwamikagami, LG 87498, LW77, La Parka Your Car, Lacis alfredo, Lacrimosus, Lankiveil, Laserforce, Lasvegas90, Laufirio, Lav150390, Lav90, Lectonar, Lenneth, Leo Plaw, Leo peter even, Leuko, Liam Powell, LichYoshi, Lightmouse, LilHelpa, Lilywatanabe, Lochdog11, Longbow4u, Longhair, Loolol, Looxix, Lotje, Luna Santin, Luong, M i k e y 86, MECU, MJCdetroit, MK8, MStraw, MagpieShooter, Maheemaravind, Maktaba, Malcolmj, Mamute41, Mandarax, Manderiko, Mangostar, Manly Regacho, Manoj Prajwal, Manticore, Marco polo, Marco79, Marcusmax, Marek69, Mark, Martyman, Mathieumcguire, Matt ke, Matthew238, Mattinbgn, Mattrix1994, Mausy5043, Maxim, Mayur, Mboys, McBanjo, Mclovin, Mecaenas, Mediatetheconflict, Mellery, Melody, Merbabu, MetroSouth, Michael Devore, Michael Johnson, Michellecrisp, Mickyounger, Midgley, Midway, Mikanmikanmikan, MikeLynch, Mikeh, Mincom2007, Mintie, Mitsuhirato, Mkidson, Mlaffs, Modulatum, Moeron, Mokgen, Moondyne, Moralis, Moreschi, Mori Riyo, Mozasaur, Mr Adequate, Mrrogerssupreme, Mrs.EasterBunny, Mrubcic2, Mufka, Mush2006, Mushroom, Muttster, Mvjs, MyFavco, MyallR, Myrtone86, NMarkRoberts, Nadermann0016, Nankai, NawlinWiki, Nebuchadnezzar o'neill, Necie B, Neep, Neiljamesmcgregor, NewEnglandYankee, Newbyguesses, Nichalp, Nicholas Perkins, Nichtayl, Nickad, Nicolemhob, Nielsen, Nikko05, Nippysaurus, Nkayesmith, Nko22, NorthernFire, Nuiop729, Nurg, NyanCat9001, Nygorilla, Nyttend, O1ive, OZ Eniak, OZTENNIS, Obradovic Goran, Octopug, Odddmonster, Ohconfucius, Ohthelameness, Oliverparish, Olivier, Orange Suede Sofa, Orderinchaos, Orderinchaos 2, Originalwana, Orodreth, OwenBlacker, P.4.P. No. 1, PDH, Packhorse, Paddington62, Papa November, PaulKidd, Pb30, Pcpp, Pdfpdf, Pee Tern, Peripitus, Peter Blake, Peter Eedy, PeterMarquisKyle, Peterphillips, Pgk, PhilAuckland, Philip Trueman, Plasma east, PoliticalGoof, Posterweb, Poz11, Princess Maria, Prolog, Proofreader77, Pulv, Pursey, Pythonlover818, Qaanaaq, Qwerty1939, Ral315, Ramisatwa, Randwicked, Raneze, RattusMaximus, Rawlins101, Rayjoe, Rbarge, Rdsmith4, Real arty, Rebecca, Recognizance, Recurring dreams, ReeseM, Rhatsa26X, RiSkE, Rich Farmbrough, Richard Keatinge, Ricky81682, RightCowLeftCoast, Rimmeraj, Riumplus, Rjwilmsi, Rlandmann, Robert Brockway, Robert Merkel, Rocketrod1960, Rodney Topor, Roger.black, RohanIsAwesome, Roidhrigh, Roke, Roland2, RoslinCastle, RowanPatterson, Rpm, Rumiton, Russavia, Ruszewski, RyanGerbil10, SEWilco, SJB689, SJP, SM247, SMC, Sakha from USA, Sakisg088, Salvio giuliano, Samulili, Saruman-the-white, Saule sauliite, SauliH, Sb617, Scene66, SchuminWeb, ScottDavis, Sean WI, Seb26, Sedination, Selket, Sendai2ci, Seoidau, Ser Amantio di Nicolao, Serein (renamed because of SUL), Shakeyourbooty2007, Shalom Yechiel, ShaneKing, Sharkface217, ShaunES, Shiftchange, ShipWatcher, Shirt58, Shootthemagpies, Shuipzv3, SidP, Silver drake, Simeon24601, Single16+Sections, SirChan, Sixers33, Sjakkalle, Sjsharksrs, Sjö, Skilldirect, Skin23, SkwEngl1101, Sl1982, Slawojarek, Sloonark, Slowking Man, Smadge1, Smalljim, Smashery, Smickel, SockGap, Soulvisionq1, Spencer, Spinrad, Spy007au, Squoups, StAnselm, Starbois, SteveJaew888, Sting au, Subtropical-man, Suffusion of Yellow, SuperHamster, Suzerain, Sweetsnacky, Swilson86, T mc 00, THEN WHO WAS PHONE?, TPK, TUF-KAT, Tabletop, Tall copy, Tamara1503, Tancred, Tannin, Tarkovsky, Tasos90, TastyPoutine, Taz300zx, Tbhotch, The Brain of Morbius, The Cat, The Famous Movie Director, The Thing That Should Not Be, The-Pope, Thebgl, Thejesterx, Thingg, Thryduulf, TicketMan, Tiffachu, Tim Starling, Time to see, Timeshift9, Tirin, Titoxd, Tobby72, Todijo, Tomaldo, Tony1, TonyTobin, Torcher382, Totoblue, Touranaut, Treff89, Trek011, Trilapoom, TripleThree, Tristan 753, Trolleymusic, Truthanado, Tuddy, Tyeby, Tyghs, Tyw7, UnderscoreGod, Utcursch, Valley2city, VanillaBear23, Vegas949, Velella, Verloren, Vianello, Vl'hurg, Vox latina, Vsquared42, WackadooXanadu, Wai Hong, Walshy93, Wavehunter, Wavelength, Wdyoung, Weate, Webhat, Welsh, Wereon, WhisperToMe, Whitehatnetizen, Widetree45, WikHead, Wikiuser95, Wm, Wolfsumatra, Wolftalker, Wongm, Woodford52, Woohookitty, Worthawholebean, Wraith61, Wraithcraze, Wriggleybum, Wtmitchell, Xabian40409, Xcentaur, Xiahou, YSSYguy, YUL89YYZ, Yamamoto Ichiro, Yameogo, YellowMonkey, Yintan, Yoyoyo11111, YuMaNuMa, Zaslav, Zero1328, ZhaoHong, Ziggurat, Zoe, Zondor, Zsinj, Zzuuzz, Александър, 1931 anonymous edits

Creative_Commons_licenses *Source*: http://en.wikipedia.org/w/index.php?title=Creative_Commons_licenses *Contributors*: A More Perfect Onion, AThing, Alansohn, Alerante, AliceUnderground, Alton, Andrew Hampe, Andromeda, Andy Dingley, Angela, Angr, Argento, Arpitt, Ashawley, AstareGod, Azrael Nightwalker, Bdonlan, BenB4, Bender235, Bennylin, CapnZapp, Chealer, Chensiyuan, Commander Keane, Creativelycommon, Cybercobra, Damian Yerrick, Damunzy, Danntm, DataWraith, Dcfleck, Dendodge, Derbeth, Deshone, DocWatson42, Dodoïste, Doldrums, Dondegroovily, Doradus, Draconx, DroEsperanto, EhJJ, Elkman, Ellmist, Eloquence, Erusse estelinya, Farmer21, FatalError, Fletcher, Flockmeal, Fornitani, FrankTobia, Frap, Gabbe, Galiaoffri, Goto, Graham france, Graue, Gronky, Guyjohnston, HaeB, Hugozam, IDGC, InMemoriamLuangPu, InverseHypercube, Isilanes, IvanLanin, J.delanoy, JSarek, JWSchmidt, Jacob Poon, Jen1979, Joaocastro, Jpatokal, Jtneill, Julian Mendez, KJG2007, Ka Faraq Gatri, Kalashnikitty, Kaldari, Kaplanmyrth, Kathleen.wright5, Ken-rene` lunding, Kl4m-AWB, Klare Kante, Kozuch, Krinkle, Kwi, LGagnon, Lars Washington, LiDEL, List yusuf, LonewackoDotCom, M7, MC10, MK8, Maierstrahl, Manscher, Martijn Hoekstra, Mauro Bieg, Mcld, Mdkoch84, Medic463, Meno25, Mgiganteus1, Mike Linksvayer, Miserlou, Mozzyepic24, MrOllie, Nemilar, NuclearWarfare, Oconnor663, Octahedron80, Oliphaunt, Omegatron, Pengo, Pgan002, Phoenix-forgotten, PullUpYourSocks, Puramyun31, Ram-Man, Rat at WikiFur, Raylopez99, Reinyday, Rich Farmbrough, RockyMM, RossPatterson, Rotem Dan, Rspeer, S401, Saba84, Sandstein, Sango123, SasiSasi, Scales, Schneelocke, Scientus, Screamingfingers, Secretlondon, Seren-dipper, Shawnc, Shinemercy, Silensor, Singlemaltscotch, Skomorokh, Smartyhall, Snek01, Srleffler, Stephenjudge, SuperLuigi31, Superm401, Synchrite, TakuyaMurata, TarzanASG, TheDoctor10, Thumperward, Torbinsky, Toussaint, Trevj, Uncle G, UpstateNYer, V111P, Vikasatkin, Vinhtantran, ViperSnake151, Volt4ire, Waldir, WelshWitch66, Wesn, Wik, Wizardist, Wuhwuzdat, XP1, Xelgen, Zondor, دمح‌ا, 78 anonymous edits

Blimp *Source*: http://en.wikipedia.org/w/index.php?title=Blimp *Contributors*: 5 albert square, 84user, A. B., A8UDI, Aannaanndd, Aarchiba, Abductive, ActivExpression, AdguruRo, Aeros40D, Airshipman, Ajstov, Akradecki, Alansohn, Alpha 4615, Amorymeltzer, Amplitude101, AnOddName, Andy Dingley, Angusmclellan, Anthony22, Apollo, Arunsingh16, BRG, Bauple58, Bensonyeung1213, BeteNoir, Black Kite, Blanchardb, Blimpguy, Bobo192, Bohuiginn, Bongwarrior, Brenth16, Brian Crawford, Brianga, Brougham96, Calliopejen1, Capricorn42, Carnildo, Cast, Celestianpower, Centrx, Chisigm321, Chuckiesdad, Church of emacs, Chzz, ColdFusion650, Creidieki, D. Recorder, Danielign9, Daveswagon, DavidLevinson, Dekisugi, Diberri, Dmsar, DocWatson42, Drmies, Drpixie, ESkog, EXTREMUS, Edward, Egil, Eloody3, Enoch Lai, Epictatus, Everyme, EvilPimppp, Ewlyahoocom, Extransit, FKrizanac, Feydey, Finngall, Fir0002, Foldoron, Football82ryan, Frecklefoot, Freebird0758, FreplySpang, Garble29, Gary D, Gary King, Giants27, Gogo Dodo, Greyengine5, Hadhuey, Haroldplace, Hiheyyoman, Hux, JGHowes, Jackehammond, Jmulder371, Jondel, Juliancolton, Junglerot56, Jusdafax, Jzhang, KPAsucxs, KaragouniS, Kcordina, Keithp41, Ken Gallager, Kickyandfun, King Lopez, KrakatoaKatie, Kubanczyk, La Pianista, Liftarn, Lommer, Lonbordin, Lothar von Richthofen, Lymantria, Lynbarn, M-le-mot-dit, MHBianchi, Marksolo, Marshaul, Martarius, Mdhennessey, MegaSloth, Mendel, Mentifisto, Mfield, Miaow Miaow, Michael Hardy, Mikelr, Mion, Montrealais, Monty845, Mpdus, Mrzaius, Mulad, Mygerardromance, Naddy, Nautical, NawlinWiki, Nickj, Nikkimaria, Nitroknight, Nsaa, Nv8200p, Nyfan1229, Olga Doktorova, Oliver202, Olivier, OllieFury, Only, Ortolan88, Paul Richter, Penubag, PeterWD, Pfalstad, Philip Trueman, Pinethicket, Pinot, Pixi Uno, Pkuxulu, Postrach, Pretzelpaws, Quistnix, Rabbit994, Red Sunset, Redjar, Rei, Rlandmann, Rodeosmurf, Rogerthat, Rory096, S!egfried, SCFilm29, SGPyro, Salvio giuliano, Sam Blacketer, Saopaulo1, Sardanaphalus, Sauvastika, SchuminWeb, Scottanon, Securiger, Skarebo, Snowaustin12, Sorruno, Station1, Stephen C. Carlson, Steven askew, Suntzu3500, Swid, SwisterTwister, Tacoman1213, Template namespace initialisation script, Tenmei, The High Fin Sperm Whale, The Thing That Should Not Be, TheMadBaron, TraceyR, Trusilver, Ttony21, Uris, VenoM5478, WOSlinker, Wahkeenah, Wedesoft, Wfaxon, Whitepaw, Wikipelli, William Leadford, Wisdom89, Woohookitty, Yeahboy179, Yworo, Zanaq, Zeimusu, Zhou Yu, Zippanova, 283 anonymous edits

Image Sources, Licenses and Contributors

File:NearMap Logo.png *Source*: http://en.wikipedia.org/w/index.php?title=File:NearMap_Logo.png *License*: unknown *Contributors*: Lakeyboy

File:Increase2.svg *Source*: http://en.wikipedia.org/w/index.php?title=File:Increase2.svg *License*: unknown *Contributors*: Sarang

File:Trapped woman on a car roof during flash flooding in Toowoomba 2.jpg *Source*: http://en.wikipedia.org/w/index.php?title=File:Trapped_woman_on_a_car_roof_during_flash_flooding_in_Toowoomba_2.jpg *License*: unknown *Contributors*: Kingbob86 (Timothy)

File:Dec 2010 Queensland floods.jpg *Source*: http://en.wikipedia.org/w/index.php?title=File:Dec_2010_Queensland_floods.jpg *License*: unknown *Contributors*: NASA. Original uploader was Peaceworld111 at en.wikipedia

File:Rockhampton in flood 4.jpg *Source*: http://en.wikipedia.org/w/index.php?title=File:Rockhampton_in_flood_4.jpg *License*: unknown *Contributors*: Tatiana Gerus from Brisbane, Australia

File:2010 Flooding in Rockhampton Queensland.jpg *Source*: http://en.wikipedia.org/w/index.php?title=File:2010_Flooding_in_Rockhampton_Queensland.jpg *License*: unknown *Contributors*: Jesse Allen

File:Burnett River in flood2.jpg *Source*: http://en.wikipedia.org/w/index.php?title=File:Burnett_River_in_flood2.jpg *License*: unknown *Contributors*: User:Sweetbixkid

File:Churchill Drive flooded in Warwick.jpg *Source*: http://en.wikipedia.org/w/index.php?title=File:Churchill_Drive_flooded_in_Warwick.jpg *License*: unknown *Contributors*: David Marsh

File:Long and Mackenzie Streets in Toowoomba flooded.jpg *Source*: http://en.wikipedia.org/w/index.php?title=File:Long_and_Mackenzie_Streets_in_Toowoomba_flooded.jpg *License*: unknown *Contributors*: Timothy

File:Wheel of brisbane during 2011 flooding.jpg *Source*: http://en.wikipedia.org/w/index.php?title=File:Wheel_of_brisbane_during_2011_flooding.jpg *License*: unknown *Contributors*: PMBO

File:2011 Flooding in Brisbane Suburbs.JPG *Source*: http://en.wikipedia.org/w/index.php?title=File:2011_Flooding_in_Brisbane_Suburbs.JPG *License*: unknown *Contributors*: ISS Expedition 26 crew

File:Ipswich_CBD.jpg *Source*: http://en.wikipedia.org/w/index.php?title=File:Ipswich_CBD.jpg *License*: unknown *Contributors*: User:Clarebear

File:Ipswich Railway Station 1865.jpg *Source*: http://en.wikipedia.org/w/index.php?title=File:Ipswich_Railway_Station_1865.jpg *License*: unknown *Contributors*: Unidentified

File:Bremer River, Ipswich, 1872.jpg *Source*: http://en.wikipedia.org/w/index.php?title=File:Bremer_River,_Ipswich,_1872.jpg *License*: unknown *Contributors*: Diglibs2, John Vandenberg, Mattinbgn, MrPanyGoff, Rwav, Soozle

File:Brisbane-street-ipswich-r.jpg *Source*: http://en.wikipedia.org/w/index.php?title=File:Brisbane-street-ipswich-r.jpg *License*: unknown *Contributors*: Digitised by Queensland Museum

File:Ipswich QLD.JPG *Source*: http://en.wikipedia.org/w/index.php?title=File:Ipswich_QLD.JPG *License*: unknown *Contributors*: User:Cgoodwin

File:Ipswich QLD (4).JPG *Source*: http://en.wikipedia.org/w/index.php?title=File:Ipswich_QLD_(4).JPG *License*: unknown *Contributors*: User:Cgoodwin

File:QRIpswich.jpg *Source*: http://en.wikipedia.org/w/index.php?title=File:QRIpswich.jpg *License*: unknown *Contributors*: Original uploader was Qld matt at en.wikipedia

Image:Goodna Jacaranda Festival Sign.jpg *Source*: http://en.wikipedia.org/w/index.php?title=File:Goodna_Jacaranda_Festival_Sign.jpg *License*: unknown *Contributors*: Goodstone

Image:Openstreetmap logo.svg *Source*: http://en.wikipedia.org/w/index.php?title=File:Openstreetmap_logo.svg *License*: unknown *Contributors*: OpenStreetMap

Image:Cambridge-Openstreetmap-11-05-27.svg *Source*: http://en.wikipedia.org/w/index.php?title=File:Cambridge-Openstreetmap-11-05-27.svg *License*: unknown *Contributors*: User:Rovastar

File:Steve Coast - OSM im Rheinland (0604).jpg *Source*: http://en.wikipedia.org/w/index.php?title=File:Steve_Coast_-_OSM_im_Rheinland_(0604).jpg *License*: unknown *Contributors*: User:Raymond

File:Openstreetmap usage stats for February 2011.png *Source*: http://en.wikipedia.org/w/index.php?title=File:Openstreetmap_usage_stats_for_February_2011.png *License*: unknown *Contributors*: Das steinerne Herz, Liftarn

Image:Mapping Strasbourg.jpg *Source*: http://en.wikipedia.org/w/index.php?title=File:Mapping_Strasbourg.jpg *License*: unknown *Contributors*: francois

Image:OpenStreetMapTracklogsHedgeEnd.png *Source*: http://en.wikipedia.org/w/index.php?title=File:OpenStreetMapTracklogsHedgeEnd.png *License*: unknown *Contributors*: Iainf, NotFromUtrecht, Ojw, 5 anonymous edits

File:OpenStreetMap homepage.png *Source*: http://en.wikipedia.org/w/index.php?title=File:OpenStreetMap_homepage.png *License*: unknown *Contributors*: User:Dream out loud

File:OpenSeaMap-Warnemünde.png *Source*: http://en.wikipedia.org/w/index.php?title=File:OpenSeaMap-Warnemünde.png *License*: unknown *Contributors*: die Mitarbeiter der beteiligten Projekte. Original uploader was Markus Bärlocher at de.wikipedia

Image:Soho - map 1.png *Source*: http://en.wikipedia.org/w/index.php?title=File:Soho_-_map_1.png *License*: unknown *Contributors*: Edward Betts

Image:JOSM-ss.png *Source*: http://en.wikipedia.org/w/index.php?title=File:JOSM-ss.png *License*: unknown *Contributors*: JOSM was written by Immanuel Scholz and other contributors

Image:Merkaartor sample.jpg *Source*: http://en.wikipedia.org/w/index.php?title=File:Merkaartor_sample.jpg *License*: unknown *Contributors*: User:Koying

Image:royal.crescent.aerial.bath.arp.jpg *Source*: http://en.wikipedia.org/w/index.php?title=File:Royal.crescent.aerial.bath.arp.jpg *License*: unknown *Contributors*: Arpingstone, Christophe.Finot, Kurpfalzbilder.de, Mahlum, Oneblackline, Ratarsed, Yarl

Image:London Docklands in Fog.jpg *Source*: http://en.wikipedia.org/w/index.php?title=File:London_Docklands_in_Fog.jpg *License*: unknown *Contributors*: User:Nuty350

File:Brooklyn Museum - Nadar Élevant la Photographie à la Hauteur de l'Art - Honoré Daumier.jpg *Source*: http://en.wikipedia.org/w/index.php?title=File:Brooklyn_Museum_-_Nadar_Élevant_la_Photographie_à_la_Hauteur_de_l'Art_-_Honoré_Daumier.jpg *License*: unknown *Contributors*: Darwinius, Zolo

Image:Balloonreflection.jpg *Source*: http://en.wikipedia.org/w/index.php?title=File:Balloonreflection.jpg *License*: unknown *Contributors*: Original uploader was SGGH at en.wikipedia

Image:Spelterini Pyramids.jpg *Source*: http://en.wikipedia.org/w/index.php?title=File:Spelterini_Pyramids.jpg *License*: unknown *Contributors*: Berrucomons, Cwbm (commons), Docu, GDK, JMCC1, JovanCormac, Lupo, Neithsabes, Yuriy75, 3 anonymous edits

File:Ilha do Fogo aerial shot 1.jpg *Source*: http://en.wikipedia.org/w/index.php?title=File:Ilha_do_Fogo_aerial_shot_1.jpg *License*: unknown *Contributors*: Aldo Bien

Image:Mtm-05277e 3d.png *Source*: http://en.wikipedia.org/w/index.php?title=File:Mtm-05277e_3d.png *License*: unknown *Contributors*: Aarchiba, ComputerHotline, El., Ruslik0

File:DTM DSM.png *Source*: http://en.wikipedia.org/w/index.php?title=File:DTM_DSM.png *License*: unknown *Contributors*: User:MartinOver

Image:Maps-for-free Sierra Nevada.png *Source*: http://en.wikipedia.org/w/index.php?title=File:Maps-for-free_Sierra_Nevada.png *License*: unknown *Contributors*: User:W!B:

File:Bezmiechowa DSM 3D 2010-05-29 Pteryx UAV.jpg *Source*: http://en.wikipedia.org/w/index.php?title=File:Bezmiechowa_DSM_3D_2010-05-29_Pteryx_UAV.jpg *License*: unknown *Contributors*: User:Kbosak

File:DSM construction site.jpg *Source*: http://en.wikipedia.org/w/index.php?title=File:DSM_construction_site.jpg *License*: unknown *Contributors*: User:Kbosak

File:Brisbane from Kangaroo Point.jpg *Source*: http://en.wikipedia.org/w/index.php?title=File:Brisbane_from_Kangaroo_Point.jpg *License*: unknown *Contributors*: User:EzykronHD

file:Australia location map.svg *Source*: http://en.wikipedia.org/w/index.php?title=File:Australia_location_map.svg *License*: unknown *Contributors*: User:NordNordWest

File:Red pog.svg *Source*: http://en.wikipedia.org/w/index.php?title=File:Red_pog.svg *License*: unknown *Contributors*: Anomie

File:Loudspeaker.svg *Source*: http://en.wikipedia.org/w/index.php?title=File:Loudspeaker.svg *License*: unknown *Contributors*: Bayo, Gmaxwell, Husky, Iamunknown, Mirithing, Myself488, Nethac DIU, Omegatron, Rocket000, The Evil IP address, Wouterhagens, 18 anonymous edits

File:StateLibQld 1 126407 R.A.A.F. recruits marching along Queen Street, Brisbane, during World War II.jpg *Source*: http://en.wikipedia.org/w/index.php?title=File:StateLibQld_1_126407_R.A.A.F._recruits_marching_along_Queen_Street,_Brisbane,_during_World_War_II.jpg *License*: unknown *Contributors*: Elekhh, Rcbutcher

File:The-Windmill-1.JPG *Source*: http://en.wikipedia.org/w/index.php?title=File:The-Windmill-1.JPG *License*: unknown *Contributors*: Original uploader was Figaro at en.wikipedia

File:ANZAC Square Brisbane.jpg *Source*: http://en.wikipedia.org/w/index.php?title=File:ANZAC_Square_Brisbane.jpg *License*: unknown *Contributors*: Bidgee, Crusoe8181, Diglibs2, Elekhh, John Vandenberg, MrPanyGoff, Rcbutcher, Shiftchange, Yarl, 2 anonymous edits

File:StateLibQld 1 104516 American fleet marching down Queen Street, Brisbane, March 1941.jpg *Source*: http://en.wikipedia.org/w/index.php?title=File:StateLibQld_1_104516_American_fleet_marching_down_Queen_Street,_Brisbane,_March_1941.jpg *License*: unknown *Contributors*: Docu, John Vandenberg

File:Expo88-from-the-Brisbane River.jpg *Source*: http://en.wikipedia.org/w/index.php?title=File:Expo88-from-the-Brisbane_River.jpg *License*: unknown *Contributors*: Original uploader was Figaro at en.wikipedia

File:First-Queenslande-Government-House-2.jpg *Source*: http://en.wikipedia.org/w/index.php?title=File:First-Queenslande-Government-House-2.jpg *License*: unknown *Contributors*: Original uploader was Figaro at en.wikipedia

File:Brisbane skyline bluesky.jpg *Source*: http://en.wikipedia.org/w/index.php?title=File:Brisbane_skyline_bluesky.jpg *License*: unknown *Contributors*: Lankiveil, Peripitus, Riana, Rjd0060, Shiftchange

File:Brisbane Aerial From Satellite.jpg *Source*: http://en.wikipedia.org/w/index.php?title=File:Brisbane_Aerial_From_Satellite.jpg *License*: unknown *Contributors*: NASA Original uploader was Lav90 at en.wikipedia

File:Brisbane CBD at Night.jpg *Source*: http://en.wikipedia.org/w/index.php?title=File:Brisbane_CBD_at_Night.jpg *License*: unknown *Contributors*: User:EzykronHD

File:Brisbane form Hamilton 2009.JPG *Source*: http://en.wikipedia.org/w/index.php?title=File:Brisbane_form_Hamilton_2009.JPG *License*: unknown *Contributors*: User:Lav90

Image:1 white, red rounded rectangle.svg *Source*: http://en.wikipedia.org/w/index.php?title=File:1_white,_red_rounded_rectangle.svg *License*: unknown *Contributors*: User:Pee Tern

Image:2 white, red rounded rectangle.svg *Source*: http://en.wikipedia.org/w/index.php?title=File:2_white,_red_rounded_rectangle.svg *License*: unknown *Contributors*: User:Pee Tern

Image:3 white, red rounded rectangle.svg *Source*: http://en.wikipedia.org/w/index.php?title=File:3_white,_red_rounded_rectangle.svg *License*: unknown *Contributors*: User:Pee Tern

Image:4 white, red rounded rectangle.svg *Source*: http://en.wikipedia.org/w/index.php?title=File:4_white,_red_rounded_rectangle.svg *License*: unknown *Contributors*: User:Pee Tern

Image:5 white, red rounded rectangle.svg *Source*: http://en.wikipedia.org/w/index.php?title=File:5_white,_red_rounded_rectangle.svg *License*: unknown *Contributors*: User:Pee Tern

Image:6 white, red rounded rectangle.svg *Source*: http://en.wikipedia.org/w/index.php?title=File:6_white,_red_rounded_rectangle.svg *License*: unknown *Contributors*: User:Pee Tern

Image:7 white, red rounded rectangle.svg *Source*: http://en.wikipedia.org/w/index.php?title=File:7_white,_red_rounded_rectangle.svg *License*: unknown *Contributors*: User:Pee Tern

Image:8 white, red rounded rectangle.svg *Source*: http://en.wikipedia.org/w/index.php?title=File:8_white,_red_rounded_rectangle.svg *License*: unknown *Contributors*: User:Pee Tern

Image:9 white, red rounded rectangle.svg *Source*: http://en.wikipedia.org/w/index.php?title=File:9_white,_red_rounded_rectangle.svg *License*: unknown *Contributors*: User:Pee Tern

Image:10 white, blue rounded rectangle.svg *Source*: http://en.wikipedia.org/w/index.php?title=File:10_white,_blue_rounded_rectangle.svg *License*: unknown *Contributors*: User:Pee Tern

Image:11 white, blue rounded rectangle.svg *Source*: http://en.wikipedia.org/w/index.php?title=File:11_white,_blue_rounded_rectangle.svg *License*: unknown *Contributors*: User:Pee Tern

Image:12 white, blue rounded rectangle.svg *Source*: http://en.wikipedia.org/w/index.php?title=File:12_white,_blue_rounded_rectangle.svg *License*: unknown *Contributors*: User:Pee Tern

Image:13 white, red rounded rectangle.svg *Source*: http://en.wikipedia.org/w/index.php?title=File:13_white,_red_rounded_rectangle.svg *License*: unknown *Contributors*: User:Pee Tern

Image:14 white, red rounded rectangle.svg *Source*: http://en.wikipedia.org/w/index.php?title=File:14_white,_red_rounded_rectangle.svg *License*: unknown *Contributors*: User:Pee Tern

Image:15 white, green rounded rectangle.svg *Source*: http://en.wikipedia.org/w/index.php?title=File:15_white,_green_rounded_rectangle.svg *License*: unknown *Contributors*: User:Pee Tern

Image:16 white, red rounded rectangle.svg *Source*: http://en.wikipedia.org/w/index.php?title=File:16_white,_red_rounded_rectangle.svg *License*: unknown *Contributors*: User:Pee Tern

Image:17 white, red rounded rectangle.svg *Source*: http://en.wikipedia.org/w/index.php?title=File:17_white,_red_rounded_rectangle.svg *License*: unknown *Contributors*: User:Pee Tern

Image:18 white, green rounded rectangle.svg *Source*: http://en.wikipedia.org/w/index.php?title=File:18_white,_green_rounded_rectangle.svg *License*: unknown *Contributors*: User:Pee Tern

Image:19 white, red rounded rectangle.svg *Source*: http://en.wikipedia.org/w/index.php?title=File:19_white,_red_rounded_rectangle.svg *License*: unknown *Contributors*: User:Pee Tern

Image:20 white, red rounded rectangle.svg *Source*: http://en.wikipedia.org/w/index.php?title=File:20_white,_red_rounded_rectangle.svg *License*: unknown *Contributors*: User:Pee Tern

Image:21 white, red rounded rectangle.svg *Source*: http://en.wikipedia.org/w/index.php?title=File:21_white,_red_rounded_rectangle.svg *License*: unknown *Contributors*: User:Pee Tern

Image:22 white, red rounded rectangle.svg *Source*: http://en.wikipedia.org/w/index.php?title=File:22_white,_red_rounded_rectangle.svg *License*: unknown *Contributors*: User:Pee Tern

Image:23 white, red rounded rectangle.svg *Source*: http://en.wikipedia.org/w/index.php?title=File:23_white,_red_rounded_rectangle.svg *License*: unknown *Contributors*: User:Pee Tern

Image:24 white, green rounded rectangle.svg *Source*: http://en.wikipedia.org/w/index.php?title=File:24_white,_green_rounded_rectangle.svg *License*: unknown *Contributors*: User:Pee Tern

Image:25 white, green rounded rectangle.svg *Source*: http://en.wikipedia.org/w/index.php?title=File:25_white,_green_rounded_rectangle.svg *License*: unknown *Contributors*: User:Pee Tern

Image:26 white, green rounded rectangle.svg *Source*: http://en.wikipedia.org/w/index.php?title=File:26_white,_green_rounded_rectangle.svg *License*: unknown *Contributors*: User:Pee Tern

Image:27 white, red rounded rectangle.svg *Source*: http://en.wikipedia.org/w/index.php?title=File:27_white,_red_rounded_rectangle.svg *License*: unknown *Contributors*: User:Pee Tern

Image:28 white, red rounded rectangle.svg *Source*: http://en.wikipedia.org/w/index.php?title=File:28_white,_red_rounded_rectangle.svg *License*: unknown *Contributors*: User:Pee Tern

Image:29 white, red rounded rectangle.svg *Source*: http://en.wikipedia.org/w/index.php?title=File:29_white,_red_rounded_rectangle.svg *License*: unknown *Contributors*: User:Pee Tern

Image:30 white, green rounded rectangle.svg *Source*: http://en.wikipedia.org/w/index.php?title=File:30_white,_green_rounded_rectangle.svg *License*: unknown *Contributors*: User:Pee Tern

File:Brisbane Town Hall.jpg *Source*: http://en.wikipedia.org/w/index.php?title=File:Brisbane_Town_Hall.jpg *License*: unknown *Contributors*: Marty Portier

File:QueenslandBuilding0005.jpg *Source*: http://en.wikipedia.org/w/index.php?title=File:QueenslandBuilding0005.jpg *License*: unknown *Contributors*: User:Sardaka

File:Brisbane Skyscrapers.jpg *Source*: http://en.wikipedia.org/w/index.php?title=File:Brisbane_Skyscrapers.jpg *License*: unknown *Contributors*: User:EzykronHD

File:FishIs.png *Source*: http://en.wikipedia.org/w/index.php?title=File:FishIs.png *License*: unknown *Contributors*: Aliasd

File:QueenStMall.jpg *Source*: http://en.wikipedia.org/w/index.php?title=File:QueenStMall.jpg *License*: unknown *Contributors*: Original uploader was James Pinnell at en.wikipedia Later version(s) were uploaded by Lilywatanabe, Figaro at en.wikipedia.

File:China-town-brisbane.jpg *Source*: http://en.wikipedia.org/w/index.php?title=File:China-town-brisbane.jpg *License*: unknown *Contributors*: Greenupp

File:Flag of the United Kingdom.svg *Source*: http://en.wikipedia.org/w/index.php?title=File:Flag_of_the_United_Kingdom.svg *License*: unknown *Contributors*: Anomie, Mifter

File:Flag of New Zealand.svg *Source*: http://en.wikipedia.org/w/index.php?title=File:Flag_of_New_Zealand.svg *License*: unknown *Contributors*: Adambro, Arria Belli, Avenue, Bawolff, Bjankuloski06en, ButterStick, Denelson83, Donk, Duduziq, EugeneZelenko, Fred J, Fry1989, Hugh Jass, Ibagli, Jusjih, Klemen Kocjancic, Mamndassan, Mattes, Nightstallion, O, Peeperman, Poromiami, Reisio, Rfc1394, Shizhao, Tabasco, Transparent Blue, Väsk, Xufanc, Zscout370, 35 anonymous edits

File:Flag of South Africa.svg *Source*: http://en.wikipedia.org/w/index.php?title=File:Flag_of_South_Africa.svg *License*: unknown *Contributors*: Adriaan, Anime Addict AA, AnonMoos, BRUTE, Daemonic Kangaroo, Dnik, Duduziq, Dzordzm, Fry1989, Homo lupus, Jappalang, Juliancolton, Kam Solusar, Klemen Kocjancic, Klymene, Lexxyy, Mahahahaneapneap, Manuelt15, Moviedefender, NeverDoING, Ninane, Poznaniak, Przemub, SKopp, ThePCKid, ThomasPusch, Tvdm, Ultratomio, Vzb83, Zscout370, 34 anonymous edits

File:Flag of Vietnam.svg *Source*: http://en.wikipedia.org/w/index.php?title=File:Flag_of_Vietnam.svg *License*: unknown *Contributors*: user:Lưu Ly

File:Flag of the People's Republic of China.svg *Source*: http://en.wikipedia.org/w/index.php?title=File:Flag_of_the_People's_Republic_of_China.svg *License*: unknown *Contributors*: User:Denelson83, User:SKopp, User:Shizhao, User:Zscout370

File:Flag of Spain.svg *Source*: http://en.wikipedia.org/w/index.php?title=File:Flag_of_Spain.svg *License*: unknown *Contributors*: Anomie

File:Flag of the Philippines.svg *Source*: http://en.wikipedia.org/w/index.php?title=File:Flag_of_the_Philippines.svg *License*: unknown *Contributors*: Aira Cutamora

File:Flag of Germany.svg *Source*: http://en.wikipedia.org/w/index.php?title=File:Flag_of_Germany.svg *License*: unknown *Contributors*: Anomie

File:Flag of India.svg *Source*: http://en.wikipedia.org/w/index.php?title=File:Flag_of_India.svg *License*: unknown *Contributors*: Anomie, Mifter

File:Flag of the Netherlands.svg *Source*: http://en.wikipedia.org/w/index.php?title=File:Flag_of_the_Netherlands.svg *License*: unknown *Contributors*: User:Zscout370

File:Flag of Fiji.svg *Source*: http://en.wikipedia.org/w/index.php?title=File:Flag_of_Fiji.svg *License*: unknown *Contributors*: Anime Addict AA, Avala, ButterStick, Denelson83, Fred the Oyster, Fry1989, Greentubing, Herbythyme, Homo lupus, Klemen Kocjancic, Krun, Lokal Profil, Mattes, Multichill, Neq00, Nightstallion, Sam916, Urhixidur, Vzb83, 8 anonymous edits

File:Flag of Papua New Guinea.svg *Source*: http://en.wikipedia.org/w/index.php?title=File:Flag_of_Papua_New_Guinea.svg *License*: unknown *Contributors*: User:Nightstallion

File:Flag of Italy.svg *Source*: http://en.wikipedia.org/w/index.php?title=File:Flag_of_Italy.svg *License*: unknown *Contributors*: Anomie

File:Flag of Malaysia.svg *Source*: http://en.wikipedia.org/w/index.php?title=File:Flag_of_Malaysia.svg *License*: unknown *Contributors*: User:SKopp

File:Flag of the United States.svg *Source*: http://en.wikipedia.org/w/index.php?title=File:Flag_of_the_United_States.svg *License*: unknown *Contributors*: Anomie

File:Flag of Hong Kong.svg *Source*: http://en.wikipedia.org/w/index.php?title=File:Flag_of_Hong_Kong.svg *License*: unknown *Contributors*: Designed by

File:Flag of South Korea.svg *Source*: http://en.wikipedia.org/w/index.php?title=File:Flag_of_South_Korea.svg *License*: unknown *Contributors*: Various

File:Flag of Sri Lanka.svg *Source*: http://en.wikipedia.org/w/index.php?title=File:Flag_of_Sri_Lanka.svg *License*: unknown *Contributors*: Zscout370

File:University of Queensland.jpg *Source*: http://en.wikipedia.org/w/index.php?title=File:University_of_Queensland.jpg *License*: unknown *Contributors*: Original uploader was Sedination at en.wikipedia

File:Southbank Beach Fireworks Night.jpg *Source*: http://en.wikipedia.org/w/index.php?title=File:Southbank_Beach_Fireworks_Night.jpg *License*: unknown *Contributors*: Brisbane Marketing and BCEC

File:Sir Thomas Brisbane Planetarium.jpg *Source*: http://en.wikipedia.org/w/index.php?title=File:Sir_Thomas_Brisbane_Planetarium.jpg *License*: unknown *Contributors*: User:Nemti

File:Suncorp Stadium.jpg *Source*: http://en.wikipedia.org/w/index.php?title=File:Suncorp_Stadium.jpg *License*: unknown *Contributors*: Original uploader was Mdmanser at en.wikipedia

File:Tennyson Tennis Centre's Pat Rafter Arena.jpg *Source*: http://en.wikipedia.org/w/index.php?title=File:Tennyson_Tennis_Centre's_Pat_Rafter_Arena.jpg *License*: unknown *Contributors*: tripletrouble

File:Mmh-&-mch-complex.jpg *Source*: http://en.wikipedia.org/w/index.php?title=File:Mmh-&-mch-complex.jpg *License*: unknown *Contributors*: Mater Health Services

File:CityCat.jpg *Source*: http://en.wikipedia.org/w/index.php?title=File:CityCat.jpg *License*: unknown *Contributors*: Ahellwig, D'oh!, Figaro, KTo288, Kgbo, Shiftchange, Smial

File:QRPassenger Suburban Multiple Unit 265.JPG *Source*: http://en.wikipedia.org/w/index.php?title=File:QRPassenger_Suburban_Multiple_Unit_265.JPG *License*: unknown *Contributors*: Gareth

File:HoughtonHighwayEveningPeak.JPG *Source*: http://en.wikipedia.org/w/index.php?title=File:HoughtonHighwayEveningPeak.JPG *License*: unknown *Contributors*: Original uploader was Seo75 at en.wikipedia

File:Etihad B777-300ER Brisbane.jpg *Source*: http://en.wikipedia.org/w/index.php?title=File:Etihad_B777-300ER_Brisbane.jpg *License*: unknown *Contributors*: User:Kwlothrop

File:King George Square Busway Station.jpg *Source*: http://en.wikipedia.org/w/index.php?title=File:King_George_Square_Busway_Station.jpg *License*: unknown *Contributors*: User:Bilious

File:Wivenhoe Dam 2011.jpg *Source*: http://en.wikipedia.org/w/index.php?title=File:Wivenhoe_Dam_2011.jpg *License*: unknown *Contributors*: User:EzykronHD

File:Brisbane During Twilight.jpg *Source*: http://en.wikipedia.org/w/index.php?title=File:Brisbane_During_Twilight.jpg *License*: unknown *Contributors*: User:EzykronHD

File:Brisbane View from Victoria Bridge.jpg *Source*: http://en.wikipedia.org/w/index.php?title=File:Brisbane_View_from_Victoria_Bridge.jpg *License*: unknown *Contributors*: User:Entropy1963

File:William Jolly Bridge.jpg *Source*: http://en.wikipedia.org/w/index.php?title=File:William_Jolly_Bridge.jpg *License*: unknown *Contributors*: User:Bilious

File:Brisbane View from Victoria Bridge 1.jpg *Source*: http://en.wikipedia.org/w/index.php?title=File:Brisbane_View_from_Victoria_Bridge_1.jpg *License*: unknown *Contributors*: User:Entropy1963

File:2010-04 Kurilpa Bridge.jpg *Source*: http://en.wikipedia.org/w/index.php?title=File:2010-04_Kurilpa_Bridge.jpg *License*: unknown *Contributors*: User:Pulv

File:Cc.logo.circle.svg *Source*: http://en.wikipedia.org/w/index.php?title=File:Cc.logo.circle.svg *License*: unknown *Contributors*: Creative Commons

File:Cc-by new.svg *Source*: http://en.wikipedia.org/w/index.php?title=File:Cc-by_new.svg *License*: unknown *Contributors*: User:Sting

File:Cc-nc.svg *Source*: http://en.wikipedia.org/w/index.php?title=File:Cc-nc.svg *License*: unknown *Contributors*: User:Rfl

File:Cc-nd.svg *Source*: http://en.wikipedia.org/w/index.php?title=File:Cc-nd.svg *License*: unknown *Contributors*: User:Rfl

File:Cc-sa.svg *Source*: http://en.wikipedia.org/w/index.php?title=File:Cc-sa.svg *License*: unknown *Contributors*: User:Rfl

Image:Goodyear-blimp.jpg *Source*: http://en.wikipedia.org/w/index.php?title=File:Goodyear-blimp.jpg *License*: unknown *Contributors*: User:Tysto

Image:GR SK Propeller.jpg *Source*: http://en.wikipedia.org/w/index.php?title=File:GR_SK_Propeller.jpg *License*: unknown *Contributors*: EugeneZelenko, Hadhuey, Korrigan, MB-one, PeterWD, Saperaud, Zaphod

Image:Blank ship in Flight.JPG *Source*: http://en.wikipedia.org/w/index.php?title=File:Blank_ship_in_Flight.JPG *License*: unknown *Contributors*: Airshipman, Akradecki, Ioeth, Pascal.Tesson

File:DIRECTV Airship in North Carolina.jpg *Source*: http://en.wikipedia.org/w/index.php?title=File:DIRECTV_Airship_in_North_Carolina.jpg *License*: unknown *Contributors*: Gfrocky, PeterWD

File:B class blimp.jpg *Source*: http://en.wikipedia.org/w/index.php?title=File:B_class_blimp.jpg *License*: unknown *Contributors*: AnRo0002, Avron, Docu, Rlandmann, 1 anonymous edits

File:Blimps.jpg *Source*: http://en.wikipedia.org/w/index.php?title=File:Blimps.jpg *License*: unknown *Contributors*: FlickreviewR, Haroldplace, Lymantria

File:Metlife snoopy two blimp.jpg *Source*: http://en.wikipedia.org/w/index.php?title=File:Metlife_snoopy_two_blimp.jpg *License*: unknown *Contributors*: User:Mfield

MIX
Papier aus verantwortungsvollen Quellen
Paper from responsible sources
FSC® C105338

Printed by Books on Demand GmbH, Norderstedt / Germany